T0339412

GROUNDWATER ARSENIC REMEDIATION

Treatment Technology and Scale UP

GROUNDWATER ARSENIC REMEDIATION
Treatment Technology and Scale UP

By

PARIMAL PAL

Professor in Chemical Engineering,
National Institute of Technology, Durgapur

AMSTERDAM • BOSTON • HEIDELBERG • LONDON
NEW YORK • OXFORD • PARIS • SAN DIEGO
SAN FRANCISCO • SINGAPORE • SYDNEY • TOKYO
Butterworth-Heinemann is an imprint of Elsevier

Butterworth Heinemann is an imprint of Elsevier
225 Wyman Street, Waltham, MA 02451, USA
The Boulevard, Langford Lane, Kidlington, Oxford OX5 1GB, UK

Notices
Knowledge and best practice in this field are constantly changing. As new research and
experience broaden our understanding, changes in research methods, professional
practices, or medical treatment may become necessary.

Practitioners and researchers must always rely on their own experience and knowledge in
evaluating and using any information, methods, compounds, or experiments described
herein. In using such information or methods they should be mindful of their own safety and
the safety of others, including parties for whom they have a professional responsibility.

To the fullest extent of the law, neither the Publisher nor the authors, contributors,
or editors, assume any liability for any injury and/or damage to persons or property
as a matter of products liability, negligence or otherwise, or from any use or operation
of any methods, products, instructions, or ideas contained in the material herein.

Library of Congress Cataloging-in-Publication Data
A catalog record for this book is available from the Library of Congress

British Library Cataloguing in Publication Data
A catalogue record for this book is available from the British Library

For information on all Butterworth Heinemann publications
visit our website at http://store.elsevier.com/

ISBN: 978-0-12-801281-9

DEDICATION

Dedicated to the memories of my late parents

CONTENTS

ACKNOWLEDGEMENTS

The idea of writing this book was planted when Kenneth P. McCombs, at Elsevier Science approached me for the same. Thanks to his very informal and helpful approach that made the subsequent steps towards publication simple and easy to follow. It has been my great pleasure to collaborate and work with Peter Jardim, Anusha Sambamoorthy, and other editorial staff members of Elsevier Science who worked so sincerely to give the final shape to the book. Role of my doctoral and post-doctoral scholars specially Ramesh, Sankha, and Jayato is very much appreciated for maintaining the system where the work was done and for checking the long list of references. Special thanks to Sankha Chakrabortty and my other doctoral scholars for tracing many figures and preparing many tables. Thanks are also due to all authors and their publishers for kind permission for use of quite a few illustrations. My sincere thanks are also due to my colleague, Dr. Mrinal. K. Mandal for valuable support. I am thankful to Prof. T. Kumar, Director N.I.T. Durgapur for doing his best despite hindrances towards my work place peace for the first time in my career. Without inspiration of the great departed souls who shaped my life, this book could not see the light of the day. Patience, tolerance and support of my dear and near ones including my brothers, sisters, wife and daughter deserve special mention.

P. Pal

PREFACE

Since the seventies of the 20th century, problems of contamination of groundwater by leached out arsenic started manifesting itself. In the nineties of the last century, reports poured over from several parts of the world such as South West USA to the Far East on acute arsenic poisoning cases. Fifty five million people in the Bengal-Delta Basin spreading over Bangladesh and India alone became victims of arsenic-contaminated groundwater. Considering widespread occurrence of arsenic and its huge carcinogenic potential, the World Health Organization set 10 ppb as the safe limit of concentration level of arsenic in drinking water in the backdrop of occurrence of as high 3000 ppb of arsenic in groundwater in many areas. Bringing down concentration of arsenic from such a high level to 10 ppb level became a challenging task to the scientific community as well as the concerned governments, and policy makers. Because of the very insidious nature of arsenic contamination problem, it takes long years for manifestation of the related diseases when the victim almost reaches a point of no return. Early government policies were mostly directed at arranging alternate sources of water such as river water by laying hundreds of kilometers of pipelines involving huge capital cost and inviting other water-borne diseases as river water sources were in many cases heavily polluted by pathogens, chemical and xenobiotic materials.

As a college student, I came across hundreds of newspaper reports and journal papers on occurrence, causes and adverse health effects of arsenic poisoning in those days. Such reports frustrated me at one point as very little on the solution front was really being talked of. I started working on the very well-known methods like adsorption and physico-chemical separation to get first-hand knowledge of the hurdles that were standing in the way of commissioning treatment plants for arsenic-bearing water. With the advent of membranes, particularly tailor-made membranes, new avenues opened up to the chemical engineering profession in the major task of separation and purification. Gradually we as a team moved into the areas of separation of arsenic by membrane-based modern techniques. We discovered that scale-up confidence in commissioning a modern arsenic treatment plant is extremely limited and this stands in a big way in implementation of modern arsenic abatement technologies. I decided to compile all our work on arsenic abatement spanning over the last two decades in to a single comprehensive book to serve as a guideline in planning, developing, commissioning, operating and maintaining on a sustainable basis a modern, efficient arsenic treatment plant. Existing books on arsenic are either on a specific area only such as occurrence or health effects or on geological explanations or at the most an edited volume containing a review of the problem and at the most partial solutions. There are now a good number of texts on membranes and membrane technology. Though these books, reviews and journal papers have enlightened us enough with which we could see better, there is hardly any book that integrates knowledge of occurrence of arsenic with the knowledge of abatement by membrane technology. That scale-up confidence is extremely limited is evident from practically non-existence of a membrane-based treatment plant in the world's most severely affected region. This book fills this gap embracing the basic understanding of the problem of occurrence of arsenic and the major technology options with their associated merits and demerits. The novel membrane distillation technique has been elaborated in the light of detailed mass and heat transfer phenomena as low flux is the major hurdle in application of membrane distillation and the same can be overcome only by improving mass and heat transfer efficiency. As standard traditional methods, adsorption and chemical coagulation have been most widely studied. Therefore, the basics and the associated merits and demerits have been discussed along with new developments. The major emphasis has however, been put on nanofiltration-based hybrid treatment that appears to the best solution being a sustainable one from the points of ecological integrity, economic efficiency and social equity. The complex issue of transport of ionic species through nanomembrane, the change of chemical speciation of arsenic through prior chemical conversion eventually producing safe potable water from arsenic-contaminated groundwater, stabilization of arsenic rejects paving the way for safe disposal need to be mathematically captured for modelling the total system for successful scale-up. This is what was absent in arsenic removal literature in a comprehensive way and has been extensively covered only in this book. For sustained operation under optimum conditions with facility for continuous visual monitoring, user-friendly Visual Basic software developed by the author and his team for physico-chemical as well as nanomembrane-integrated hybrid processes is indeed new addition to the literature of arsenic abatement technology through this book. However, to help select a technology for a particular situation, understanding other major technology options is also essential. This book deals with all the broad options of treatment of arsenic-bearing water. Each major chapter deals with one technology. After the foundation has been laid out with the dissemination of knowledge on occurrence, causes, health effects, regulations and the desired level of purification, there has been logical progression of these chapters starting from the very basics of the science and technology to its gradual development towards practical solution involving modelling, simulation, optimization, control, economic analysis and scale-up. The greatest strength of this book lies in the firsthand experience of an author belonging to the world's worst affected arsenic zone who investigated with real arsenic-contaminated groundwater, experimentally collected data on the possible best solutions for arsenic-affected areas, designed and fabricated new devices for treatment of arsenic-bearing water and enriched himself with knowledge from the findings of the tireless research carried out by the scientists all over the world. Findings of the research activities on arsenic separation carried out in the Environment and Membrane Technology Laboratory of the Department of Chemical Engineering, National Institute of Technology Durgapur have been acknowledged by the scientific community of the world through several publications of the reputed publishers like Elsevier Science, Springer, Taylor and Francis, American Chemical Society, Water Environment Federation, USA and International Water Works Association (IWA). Possible schemes of treatment have been investigated and have been analyzed chemically, mathematically and socio-economically before suggesting a solution. Comparisons have been made in terms of separation efficiency, complexity and cost of operation and sustainable management. The book should be able to help people arrive at a quick solution in selecting the best scheme and in scaling up, setting up and successfully operating an efficient arsenic removal plant depending on the ground realities encompassing geo-politico and socio-economic conditions. Even the issues of final disposal of concentrated arsenic rejects have received attention. Integration of so many novel ideas from diverse points of view in the context of removal of arsenic from contaminated water into a single book makes this volume unique.

Scholars from undergraduate level to research level are likely to find the book useful along with membrane suppliers, planners, governments, public health engineers, managers and policy makers entrusted with the great responsibility of providing safe potable water to the people in the affected regions.

Parimal Pal

1st January, 2015.

CHAPTER 1

Introduction to the Arsenic Contamination Problem

Contents

Contamination of groundwater by leached out arsenic results in a serious human health problem when underground aquifers are used as a source of drinking water. Millions of people in over 20 countries across the world such as Argentina, Bangladesh, China, India, Mexico, Mongolia, Nepal, Taiwan, Thailand, the United States, Canada, Vietnam, among others face this arsenic contamination problem [1–25]. In some places in Bangladesh, concentration of arsenic in groundwater is as high as 1000 μg/L. The largest population at risk among the countries affected by arsenic contamination of groundwater is in Bangladesh, followed by West Bengal in India. In offering relief to the suffering populations, region-specific policy formulation that considers the huge variation of socioeconomic conditions and availability of technology is essential. Despite extensive research on the occurrence of arsenic and on mitigation methods, implementation of arsenic removal

Groundwater Arsenic Remediation
http://dx.doi.org/10.1016/B978-0-12-801281-9.00001-1

1

technology is very limited in the backdrop of limited scale-up confidence. In most cases, arsenic-free surface water from an alternate source is arranged instead of using effective arsenic removal technology for making the contaminated water free from arsenic. But in areas where a source of alternate safe water is not available, technology for the separation of arsenic from contaminated water remains the only option. This necessitates adoption of a new approach toward solving this problem, one that meets the scientific challenge of removing arsenic from water instead of ignoring the problem. It is thus of paramount importance to understand the chemistry of arsenic in the water environment, the causes of its occurrence in groundwater, adverse health effects, and possible technological mitigation approaches.

1.1 ARSENIC CHEMISTRY

Arsenic with atomic number 33 is located in group VA of the periodic table directly below phosphorous. Because of the physico-chemical similarity with phosphorus, often ingested arsenic in the human body disrupts the ATP cycle and hence the metabolism system by replacing phosphorus. Arsenic is the 20th most abundant element in the earth's crust, 14th in seawater, and the 12th most abundant element in the human body. Arsenic forms inorganic and organic compounds and may occur in the environment in various oxidation states (-3, 0, $+3$, $+5$). In natural water, arsenic occurs mainly in inorganic forms as oxyanions of trivalent ($+3$) arsenite or as pentavalent ($+5$) arsenate. Inorganic compounds of arsenic include hydrides (e.g., arsine), halides, oxides, acids, and sulfides. Examples of inorganic arsenic oxide compounds include As_2O_3, As_2O_5, and arsenic sulfides such as As_2S_3, $HAsS_2$, and $HAsS_3^{3-}$. The two oxidation states predominant in ground and surface waters are arsenate (V) and arsenite (III) and are part of the arsenic (H_3AsO_4) and arseneous (H_3AsO_3) acid systems, respectively. Inorganic arsenic species that are stable in oxygenated waters include arsenic acid and As (V) species such as H_3AsO_4, $H_2AsO_4^{-}$, $HAsO_4^{2-}$, and AsO_4^{3-}. Arseneous acid (As(III)) is also stable as H_3AsO_3, $H_2AsO_3^{-}$, and $HAsO_3^{2-}$ under slightly reducing aqueous conditions. These two forms of arsenic depend upon oxidation–reduction potential and pH of the water [26]. Different structures of arsenic compounds are presented in Figure 1.1. At a typical pH of 5.0 to 8.0 of natural water, the predominant pentavalent arsenate species are $H_2AsO_4^{-}$ and $HAsO_4^{2-}$ and the trivalent arsenite species is H_3AsO_3. The ratio of As(v) to As (III) in natural water is about 4:1. Arsenates are stable under aerobic or oxidizing conditions, while arsenite

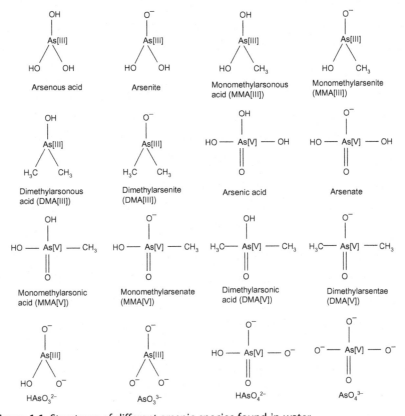

Figure 1.1 Structures of different arsenic species found in water.

compounds are stable under anaerobic or mildly reducing conditions. In reducing waters, arsenic is found primarily in the trivalent oxidation state in the form of arseneous acid that ionizes according to the following equations:

$$H_3AsO_3 \leftrightarrow H^+ + H_2AsO_3^- \qquad pKa = 9.22$$
$$H_2AsO_3^- \leftrightarrow H^+ + HAsO_3^{2-} \qquad pKa = 12.3$$

The acid–base dissociation reactions of arsenic acid can be described as:

$$H_3AsO_4 \leftrightarrow H^+ + H_2AsO_4^- \qquad pKa = 2.20$$
$$H_2AsO_4^- \leftrightarrow H^+ + HAsO_4^{2-} \qquad pKa = 6.97$$
$$HAsO_4^{2-} \leftrightarrow H^+ + AsO_4^{3-} \qquad pKa = 11.53$$

pKa is the pH at which the disassociation of the reactant is 50% complete. The dominant organic forms found in water are methyl and dimethyl arsenic compounds, such as monomethyl arseneous acid (MMA (III)), monomethyl arsenic acid (MMA(V)), dimethyl arseneous acid (DMA(III)), and dimethyl arsenic acid (DMA(V)) [27].

Arsenic speciation in aqueous solution is controlled by the two most important factors, redox potential (Eh) and pH. Figure 1.2 describes the relationships between Eh, pH, and aqueous arsenic species. Under oxidizing conditions, $HAsO_4^{2-}$ dominates at a high pH regime whereas H_3AsO_4 and AsO_4^{2-} predominate in extremely acidic and alkaline conditions, respectively (Table 1.1). $H_2AsO_4^-$ predominates at low pH (<6.9).

Under reducing conditions at a pH of less than 9.2, the uncharged species H_3AsO_3 will predominate (Figure 1.2). This means that As(III) remains as a neutral molecule in natural water. Generally, pentavalent arsenic species is the dominant arsenic species in surface water since As(V) species are stable in the oxygen-rich aerobic conditions (positive Eh value). On the other hand, trivalent arsenic is more likely to occur in groundwater since As(III) species are thermodynamically stable and dominant in mildly reducing

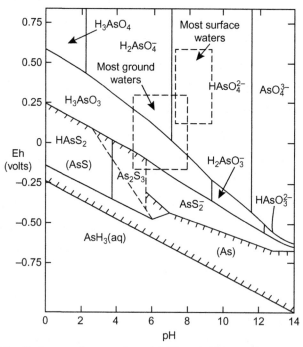

Figure 1.2 Eh-pH diagram for arsenic at 25 °C and 1 bar total pressure, with total arsenic 10^{-5} mol/L; symbols for solid species are enclosed in parentheses in crosshatched area, which indicates solubility less than 10^{-5} mol/L. *(Adapted from [26]).*

anaerobic conditions, which are the characteristics of most groundwater (negative Eh value).

The thermodynamic equilibrium diagrams for As(III) and AS(V) have been constructed with computer code [28] and are described in Figures 1.3 and 1.4.

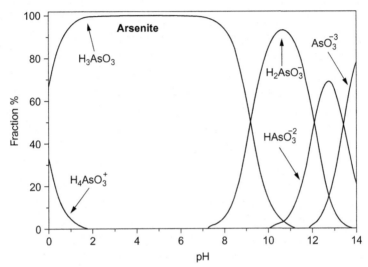

Figure 1.3 Arsenic speciation as a function of pH for total As(III) concentration 50 mg/L (constructed with computer code [28]).

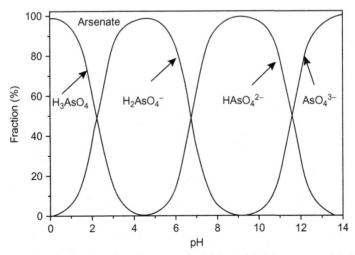

Figure 1.4 Arsenic speciation as a function of pH for total As(V) concentration 50 mg/L (constructed with computer code [28]).

1.2 OCCURRENCE AND CAUSES OF ARSENIC IN GROUNDWATER

Arsenic may occur naturally in some 200 minerals in varying degrees as elemental arsenic, arsenide, sulfides, oxides, arsenite, and arsenates. The highest concentrations of arsenic however, are associated with sulfide minerals and metal oxides, especially iron oxides. The problem of arsenic contamination can result in places of abundant occurrence of these minerals only if the geochemical conditions favor release of arsenic from these minerals. The geochemical conditions that are widely accepted as causes of groundwater contamination are pH, aerobic or reducing environment, groundwater flow, and transport. The most abundant arsenic ore mineral is pyrite (FeS_2) followed by chalcopyrite, galena, and marcasite, where arsenic concentrations can be as high as 10 w%. Besides being an important ore component, pyrite can also be formed in sedimentary environments under reducing conditions.

Under aerobic conditions, pyrite may get oxidized to iron oxides with release of sulfate, acidity, and arsenic along with trace elements. The pyrite oxidation reactions take place following the equation

$$FeS_2 + 15/4\,O_2 + 7/2\,H_2O \rightarrow Fe(OH)_3 + 2SO_4^{-2} + 4\,H^+$$

Thus human activities around coal mining often are blamed for arsenic problems in coal mine areas. Fortunately, under most circumstances, mobilization of arsenic to groundwater and surface water is low because of high retention of arsenic species in the associated minerals. Arsenic can also be released from arsenopyrite (FeSAs) under aerobic conditions and in many aquifers around the world, lowering of the water table has been held responsible for creating an aerobic environment through the introduction of atmospheric oxygen [29,30]. Another school of thought [31,32] is that arsenic leaching caused by biomediated reductive dissolution of arsenic-bearing ferric-oxyhydroxides is mainly responsible for the problem of arsenic contamination in the Bengal delta basin. Reduction of oxyhydroxides in alluvial aquifers needs organic matter (OM), the source of which may be anthropogenic (poor sanitation, surface soils) or authigenic. Though organic matter drives reduction, the surface source of such organic matter is almost ruled out [33], contrary to some observations [14] in the case of arsenic-polluted groundwater in Pakistan's Muzzafargarh. In the Bengal delta basin, organic-rich fluvio-deltaic sediments that were deposited during the high-stand setting of the mid-Holocene age [34] are found to be associated with a major arsenic contamination problem.

Authogenic sulfide minerals containing arsenic can be formed under strongly reducing conditions in lakes, oceans, and aquifers. Oxidation conditions that often cause dissolution of arsenic from sulfide minerals may happen in shallow aquifers and not in deep aquifers. However groundwater may remain in oxic conditions for 5000 years when the associated sediment itself is organic-poor, as in the Sherwood sandstone aquifer in the United Kingdom [35]. In the Terai region of Nepal the source of arsenic is believed to be geogenic [36], where arsenic contamination of groundwater is attributed to reductive dissolution of ferro-oxyhydroxide.

In controlling the redox conditions of reducing aquifers, the role of organic matter has been widely suggested [37], though there remains some dispute on the nature of organic matter. Rapid burial of organic matter along with sediments facilitates microbial activities, which generate reducing conditions favorable to the formation of sulfide minerals containing arsenic [32–37]. Sewage, animal, and human wastes (anthropogenic organic matter) also cause reduction of hydrous ferric oxide and release of sorbed arsenic into shallow underground aquifers (<30 m) though surface source of organic matter in driving such a reduction process is considered extremely unlikely [14]. However, reducing conditions in deep (>30 m) aquifers seem to be due to naturally occurring organic matter. In various parts of Asia, onset of reducing conditions in the sediment and later conditions of oxidation in the aquifers largely have been held responsible for the problem of groundwater contamination by arsenic. After release of arsenic from crystal lattice and its dissolution in water, accumulation of arsenic in the aquifer may continue unless it is flushed out by moving groundwater over time. Slow groundwater movement (due to low recharging rate) has been blamed for many high arsenic aquifers in countries in South East Asia. Thus low arsenic concentrations in deep and coastal aquifers in Bangladesh and elsewhere is attributed to high groundwater movement and high rate of recharging.

In some geologically recent and poorly flushed arid and semiarid regions of the world like the inland basins of Argentina and the southwestern United States, high pH conditions have resulted in desorption of arsenic from mineral surfaces [38]. Mineral weathering and high evaporation lead to high pH conditions. It is well established that under aerobic and low-to-neutral pH regime, adsorption of arsenic, especially as As(V) on iron oxides, is very strong, aqueous concentrations are low, and arsenic desorption is favored at high pH. The role of microbes in reduction and mobilization of arsenic has also been observed [39,40].

Many factors influence the concentration of arsenic in the natural environment, such as organic and inorganic components of the soils, redox potential status. Volcanic activity and the erosion of rocks and minerals are also sources that can release arsenic into the environment, as are anthropogenic activities. Arsenic-containing substances such as wood preservatives, paints, drugs, dyes, metals, and semiconductors may also release arsenic directly to the environment. Agricultural applications (pesticides, fertilizers), mining, smelting, land filling, and other industrial activities contribute to arsenic contamination in the environment. Arsenic present in water is due to these natural and anthropogenic activities.

Water is one of the principal means of transport of arsenic in the environment. In seawater, arsenic occurs in pentavalent, trivalent, and methylated forms. The seawater ordinarily contains 1.5 to 5 μg/L arsenic [41]. In the photic zone of seawater, a high content of the trivalent form was detected, due to the fact that pentavalent arsenic is transferred to trivalent arsenic and organo-arsenic compounds through biological activity. Under natural conditions, arsenic is predominant in places with high geothermal activities in the aquatic environment. In the second region of Chile, the streams are characterized by a high arsenic content (100–1,000 mg/L), mostly associated with the geothermal activity and quaternary volcanic in Andes Cordillera [20]. Industrial activities cause also serious pollution problems in nearby groundwater. The groundwater in Reppel (north Belgium), which is an industrial site polluted with arsenic and heavy metal, contains up to 31,000 mg/L [42].

Microbial agents can influence the oxidation state of arsenic in water, and can mediate the methylation of inorganic arsenic to form organic arsenic compounds [43]. Microorganisms can oxidize arsenite to arsenate, and reduce arsenate to arsenite or even to arsine (ASH_3). Bacteria and fungi can reduce arsenate to volatile methyl arsines. Marine algae transform arsenate into nonvolatile methylated arsenic compounds such as methyl–arsenic acid ($CH_3AsO(OH)_2$) and dimethylarsinic acid ($(CH_3)_2AsO(OH)$) in seawater. Fresh water and marine algae and aquatic plants synthesize complex lipid-soluble arsenic compounds [44]. Organic arsenical compounds were reported to have been detected in surface water more often than in ground water. Surface water samples reportedly contain low but detectable

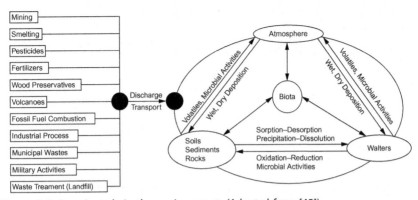

Figure 1.5 Arsenic cycle in the environment. *(Adapted from [45]).*

Table 1.1 Eh-pH Values for Arsenic Speciation in Aqueous Solution in the System As-O_2-H_2O at 25 °C and 1 Bar Total Pressure*

Species	Eh mV (approximate)		pH range (approximate)
	Maximum	Minimum	
H_3AsO_3	600 at pH$=0$	−528	Less than about 9.2
$H_2AsO_3^-$	−233	−681	Between 9.2 and 12.2
$HAsO_3^{2-}$	−509	−762	Between 12.2 and 13.5
AsO_3^{3-}	−605	−795 at pH$=14$	Above 13.5
H_3AsO_4	1224 at pH$=0$	452	Less than about 2.2
$H_2AsO_4^-$	1105	81	Between 2.2 to 6.7
$HAsO_4^{2-}$	843	−424	Between 6.7 to 11.5
AsO_4^{3-}	576	−629 at p$^H=14$	Above 11.5

*Calculated from Eh-pH diagram

concentrations of arsenic species including methyl arsenic acid and dimethylarsinic acid. Methyl-arsenicals have been reported to comprise as much as 59% of total arsenic in lake water. In some lakes, dimethylarsinic acid has been reported as the dominant species, and concentrations appear to vary seasonally as a result of biological activity within waters. Figure 1.5 shows the occurrence and flow paths of arsenic in the environment.

Arsenic has been found at higher levels in underground drinking water sources than in surface waters. Arsenic concentrations in environmental media are presented in Table 1.2.

Table 1.2 Arsenic Concentrations in Environmental Media (US-Environmental Protection Agency 2000)

Environmental media	Arsenic concentration range
Air, ng/m3	1.5–53
Rain from unpolluted ocean air, µg/L (ppb)	0.019
Rain from terrestrial air, µg/L	0.46
Rivers, µg/L	0.20–264
Lakes, µg/L	0.38–1,000
Ground (well) water, µg/L	< 1.0 and $>1,000$
Seawater, µg/L	0.15–6.0
Soil, mg/kg (ppm)	0.1–1,000
Stream/river sediment, mg/kg	5.0–4,000
Lake sediment, mg/kg	2.0–300
Igneous rock, mg/kg	0.3–113
Metamorphic rock, mg/kg	0.0–143
Sedimentary rock, mg/kg	0.1–490
Biota: green algae, mg/kg	0.5–5.0
Biota: brown algae, mg/kg	30

1.3 REGULATIONS AND MAXIMUM CONTAMINANT LEVEL OF ARSENIC

Acute and chronic arsenic exposure via drinking water has been reported in many countries, especially Argentina, Bangladesh, India, Mexico, Mongolia, Thailand, and Taiwan, where a large proportion of groundwater is contaminated with arsenic at levels from 100 to over 2,000 µg/L (ppb). The toxicity of arsenic in humans at small doses (mg/kg body weight) is well known. Within the United States, a maximum permissible concentration of 0.05 mg/L for arsenic in drinking water was first established by the Public Health Service in 1942. Beginning in 1968 studies relating arsenic exposure and skin cancer began to raise questions about the adequacy of the 0.05 mg/L standard. Throughout the 1980s and into the 1990s the Unites States Environmental Protection Agency (USEPA) attempted to establish a revised Maximum Contaminant Level (MCL) for arsenic. Strong epidemiological evidence of arsenic carcinogenicity and genotoxicity has forced the World Health Organization (WHO) to lower the maximum MCL in drinking water to 10 ppb from an earlier limit of 50 ppb in 1993, followed by the USEPA adoption of the same in 2001. However, the prescribed MCL of arsenic in drinking water (Table 1.3) is found to vary from country to country [46]. While the value is 50 ppb in developing countries like Bangladesh,

Table 1.3 Maximum Contaminant Level (MCL) of Arsenic Set by Different Countries

Countries/others	Maximum contaminant level, ppb
WHO/USEPA/European Union	10
Germany	10
Australia	7
France	15
India, Bangladesh, Vietnam, Mexico	50
Malaysia	10–50

India, China, and Taiwan, it is 10 ppb in developed countries like the United States, Germany, and Japan, 25 ppb in Canada, and 7 ppb in Australia.

1.4 TOXICITY AND HEALTH HAZARDS

Many pollutants in water streams have been identified as toxic and harmful to the environment and human health. Among them arsenic is considered a high priority. Arsenic has been identified as a Class I human carcinogen and is a public concern due to its widespread usage in both industry and agriculture. Arsenic cannot be easily destroyed and can only be converted into different forms or transformed into insoluble compounds in combination with other elements such as iron. Many impurities such as lead, iron, and selenium may be mixed together with arsenic waste which makes it uneconomical to remove. The contaminants like iron, calcium, magnesium, bicarbonate, chloride, and sulfate are found to be associated with arsenic in groundwater. The toxicology and carcinogenicity of arsenic depend on its oxidation states and chemical forms. While inorganic arsenic is more toxic than organic arsenic (except MMA(III) and DMA(III)), the trivalent form is more hazardous than the pentavalent form [47]. The pentavalent arsenic (arsenate) can replace the role and position of phosphate in the human body due to its similar structure and properties with phosphate. For example, arsenate can disrupt the formation process of high energy phosphate bonds (in ATP), a primary energy storage form in the cell. This disruption results in loss of energy. The activity of many enzymes, coenzymes, and receptors containing thiol groups (–SH) are destroyed by trivalent arsenic (arsenite) due to bonding of arsenite with the thiol group within enzymes as biological catalyst. For example, arsenite can react with liplic acid, containing dithiol groups. This reaction inhibits PDH (pyruvate dehydrogenase), which requires liplic acid for enzymatic activity. This inhibition will impede the Krebs cycle by destroying the function of PDH.

Arsenite is considered to be more toxic than arsenate. It has been reported that As(III) is 4 to 10 times more soluble in water than As(V) [48]. Moreover, it has been found that As(III) is 10 times more toxic than As(V) and 70 times more toxic than MMA(V) and DMA(V). Inorganic arsenic and organo-arsenicals like dimethylarsinic acid can lead to great genotoxicity in low and micromolar doses. However, the trivalent methylated arsenic species like MMA(III) and DMA(III) have been found to be more toxic than inorganic arsenic because they are more efficient at causing DNA breakdown [49]. The toxicity of different arsenic species varies in this order: arsenite → arsenate → mono-methyl-arsenate (MMA) → dimethylarsinate (DMA) [50]. Arsenic poisoning has become one of the major environmental worries in the world as millions of human beings have been exposed to excessive arsenic through contaminated drinking water. Toxicology and carcinogenicity of arsenic depend on its oxidation states and chemical forms. While inorganic arsenic is more toxic than organic arsenic, the trivalent form is more hazardous than the pentavalent form [47]. Many impurities such as lead, iron, and selenium may be mixed together with arsenic wastes, making it uneconomical to remove. The contaminants like iron, calcium, magnesium, bicarbonate, chloride, and sulfate are found to be associated with arsenic in the groundwater in many cases. Inorganic arsenic, As(V) and As(III) in the form of Na_2HAsO_4 and $NaAsO_2$, respectively, are toxic to humans and plants. Inorganic arsenic is always considered a potent human carcinogen. It also has noncancer effects that include cardiovascular, pulmonary, immunological, neurological, and endocrine (e.g., diabetes) disorders. Besides its tumorigenic potential, arsenic has been shown to have genotoxicity [51]. The toxicology of arsenic can be classified into acute and subacute types. The poisoning of arsenic requiring prompt medical attention usually occurs through ingestion of contaminated food or drink. The major early manifestation of acute arsenic poisoning includes burning and dryness of the mouth and throat, dysphasia, colicky abnormal pain, projectile vomiting, profuse diarrhea, and hematuria.

India (mainly the state of West Bengal) and Bangladesh have long suffered from the problem of arsenic contaminated groundwater and claims the biggest calamity in the world. The much hidden nature of groundwater poisoning by leached out arsenic surfaced in an acute form in the Bengal delta basin when hundreds of people started showing symptoms of arsenic-related diseases. Reportedly the problem manifested itself after some two decades of continuous consumption of arsenic-contaminated water [52]. Arsenic concentration of over 60 mg/L is lethal for human consumption [53].

To prevent arsenic-related diseases, the maximum contaminant level (MCL) in drinking water has been set by different countries as shown in Table 1.3.

Detailed clinical examination and investigation of 248 such patients revealed protean clinical manifestations of toxicity. Over and above hyper-pigmentation and keratosis, weakness, anemia, burning sensation of eyes, solid swelling of legs, liver fibrosis, chronic lung disease, gangrene of toes, neuropathy, and skin cancer are some of the other manifestations [54].

Naturally occurring arsenic, adsorbed from rocks through which water passes, is present in some 4,000 sites in the United States, mainly in the southwest and northeast states. Utilities supplying water complied with earlier EPA standards of a 50 μg/L maximum contaminant level, but the revised compliance levels that reduced this to 10 μg/L represented a big change. In Bangladesh, 2000 villages have been identified as containing arsenic above 50 μg/L, and over 50 million people have been exposed to arsenic poisoning. In India over 2700 villages are affected by arsenic poisoning in groundwater. Over 6 million people are consuming arsenic-contaminated water and there are over 30,000 reported cases of those already affected by arsenic [35]. The occurrence and toxicity of arsenic have been reported comprehensively [50]. The toxicology of arsenic is a complex phenomenon and generally classified as acute and subacute. Acute arsenic poisoning requires prompt medical attention. It usually occurs through ingestion of contaminated food or drink. The major early manifestation due to acute arsenic poisoning includes burning and dryness of the mouth and throat, dysphasia, colicky abnormal pain, projectile vomiting, profuse diarrhea, and hematuria. Muscular cramps, facial edema, cardiac abnormalities, and shock can develop rapidly as a result of dehydration.

In general, there are four recognized stages of arsenicosis or chronic arsenic poisoning: preclinical, clinical, complications, and malignancy. In the preclinical stage, the patient shows no symptoms, but arsenic can be detected in urine or body tissue samples. In the clinical stage, various effects can be seen on the skin. Darkening of the skin (melanosis) is the most common symptom, often observed on the palms. Dark spots on the chest, back, limbs, or gums have also been reported. Edema (swelling of hands and feet) is often seen. A more serious symptom is keratosis, or the hardening of skin into nodules, often on palms and soles. WHO estimates that this stage requires 5 to 10 years of exposure to arsenic. In the complications stage, clinical symptoms become more pronounced and internal organs are affected. Enlargement of the liver, kidneys, and spleen have been reported. Some research indicates that conjunctivitis (pinkeye), bronchitis, and diabetes may be linked to arsenic exposure at this stage. Tumors or cancers (carcinoma) affect skin or other organs in the malignancy stage. The affected person may develop gangrene or skin, lung, or bladder cancer.

The results of clinical findings for arsenic poisoning from drinking arsenic-contaminated water show the presence of almost all the stages of arsenic clinical manifestation [55]. Exposure to arsenic via drinking water (groundwater) has been reported to cause a severe disease of blood vessels leading to gangrene, known as black foot disease in Taiwan [56].

After absorption, inorganic arsenic accumulates in the liver, spleen, kidneys, lungs, and gastrointestinal tract. It is then rapidly cleared from these sites but leaves a residue in keratin-rich tissues such as skin, hair, and nails. Arsenic, particularly in its trivalent form, inhibits critical sulfhydryl-containing enzymes. In the pentavalent form, the competitive substitution of arsenic for phosphate can lead to rapid hydrolysis of the high-energy bonds in compounds such as ATP. Chronic exposure to high levels of arsenic concentration in drinking water has been associated with cancers of the skin, lung, liver, and kidney in different arsenic-affected parts of the world. Arsenic poisoning has also been blamed for several cardiovascular, cerebrovascular, endocrine-disrupting, and neurodevelopmental diseases [57–59]. Even at low-to-moderate doses of arsenic poisoning, adverse health effects in the form of premalignant skin lesions, high blood pressure, and neurological dysfunctions have been reported in the arsenic longitudinal study in Bangladesh [57]. Poor health and hygiene, relatively low affordability of the greater majority of the population living in these zones, and the lack of awareness of the possible consequences of arsenic intoxication make this problem more complex [60].

The vastness of this problem calls for a tremendous all-out effort to bring the situation under control. Chronic exposure generally leads to various ailments and the dysfunction of several vital organs like the liver, kidney, and lungs, tremor-producing effects, neurological disorders, and so on, more often when there is an accompanying nutritional/dietary deficiency [61]. Most of the affected people in general complain of muscle and joint pains and are highly depressed with various gastric problems and general weakness [24]. Skin and nail changes like arsenical dermatitis, melanosis, keratosis, gangrene, sensory and motor polyneuritis, hepatitis/chronic liver disease, chronic diarrhea, aplastic anemia, hyperostosis, portal hypertension, toxic optic neuropathy(atrophy), skin cancer, and cancers of the lung, liver (angiosarcoma), bladder, kidney, and colon [62] are found to occur. The role of antioxidants has been found to be very successful in combating the arsenic-related diseases [63,64]. Efficacy of vitamin E and selenium supplementation in treating patients with arsenic-induced skin lesions has been studied [63]. However, orthodox medicines (e.g., chelating agents like Dimercaptosuccinic acid (DMSA), Diethylenetriamine-pentaacetic acid (DTPA), etc., and some antioxidants) have been most unsuccessful. Thus besides water purification, suitable antagonists of arsenic poisoning need to be discovered that would be (1) easy to administer, (2) effective in low doses, (3) inexpensive, and (4) without any toxic effects of their own. Studies convincingly demonstrate that the potentized homeopathic drug, Arsenicum Album, not only has the ability to help removal of arsenic from the body, but in micro doses, appear to have the ability to detoxify the ill effects produced by arsenic [60]. Much evidence-based research on human beings and animals have shown the efficacy of homoeopathic drugs in combating this problem [65].

1.5 INTRODUCTION TO METHODS OF ARSENIC REMOVAL

The arsenic removal methods include precipitation, adsorption, ion exchange, coagulation and flocculation, and membrane separation. Other precipitation methods have been studied for arsenic removal using hydrogen peroxide, calcium oxide, ferric sulfate, and Portland cement as the precipitation agents.

1.5.1 Chemical Precipitation

In the precipitation process, anions combine with cations resulting in precipitation. Three processes are well known: alum coagulation, iron coagulation, and lime softening. The disadvantages of the chemical precipitation process include (1) requirement of a large amount of chemicals and generation of volumetric sludge; and (2) formation of unstable Arsenic III sulfide, calcium arsenate, or ferric arsenate precipitates under certain conditions.

1.5.1.1 Alum Precipitation

This process is effective for the removal of solids and dissolved metals. Chemicals required for the process are chlorine, acid, alum, and caustic soda. Acid is required to maintain pH at the desired level. To increase pH to an acceptable level in the posttreatment of clarified water, caustic soda (NaOH), for example, would be added. The alum sludge generated in the clarifier contains arsenic removed from the water.

1.5.1.2 Iron Precipitation

In this process, a ferric salt (for example, $FeCl_3$ and $Fe_2(SO4)_3$) and chlorine, as an oxidizing agent, are added. The arsenic combined with the iron forms a precipitate that settles out in the clarifier. The particles of iron/arsenic that are not settled out in the clarifier are removed by employing a filter, followed by a clarifier. Ninety-five percent removal of arsenic from water containing 300 μg/L of arsenic using 30 mg/L of $Fe_2(SO4)_3$ is achieved at a pH of less than 8.5 with chlorine.

1.5.1.3 Lime Softening

In this process, arsenic is removed with other particles from water other than hardness (calcium and magnesium ions). The chemicals required for the process are chlorine, lime, and acid. Chlorine is needed to oxidize the arsenic. Acid is necessary to lower the pH of the treated water to acceptable drinking water levels.

1.5.1.4 Coprecipitation

This process is applied to remove arsenic along with iron (and/or manganese) from arsenic and iron (and/or manganese) contaminated water. The principle of separation is oxidizing the iron and/or manganese from their soluble state (oxidation state 2+) to a higher oxidation state to form iron and/or manganese precipitates. The arsenic is apparently removed as iron/arsenic or manganese/arsenic precipitates, which are backwashed off of the filter media. Then the precipitates can be filtered. The most important chemical used in this process is chlorine as an oxidizing agent. Other chemicals such as ferric chloride, sulfur dioxide, potassium permanganate, and polymeric aluminum silicate sulfate (PASS) may or may not be needed, depending on the water chemistry and process employed.

1.5.2 Adsorption

The technology of adsorption uses materials that have a strong affinity for dissolved arsenic. Arsenic is attracted to the sorption site on the adsorbent's surface and is removed from water. This process is efficient for arsenic removal from drinking water. Activated carbon is the well-known widely used adsorbent. However, activated carbon still remains an expensive material.

1.5.3 Ion Exchange

Ion exchange is the process of exchanging arsenic anions for chloride or other anions at active sites bound to a resin. It is an adsorption process similar to activated alumina. An ion exchange resin, attached with chloride ions at the exchange sites, is placed in a vessel. The arsenic-containing water is passed through the resin bed and the chloride ion is exchanged by arsenic anions. The water coming out from the resin bed is lower in arsenic but higher in chloride than the water entering the vessel. When all or most of the exchange sites are occupied by arsenic or other anions by replacing chloride ions, the resin gets exhausted. The exhausted resin is regenerated with salt (sodium chloride). During the regeneration process there are substantial concentrations of sodium and chloride in the wastewater as well as the arsenic. The resins prefer sulfate ions to arsenic anions when arsenic contaminated water containing sulfate ions is treated in the ion exchange process. As a result, the sulfate ions are exchanged for chloride ions before the arsenic ions.

Ion exchange resin can only exchange anions from water; that is, arsenic in the form of anions can only replace chloride ions attached to the resin. The pH of the feed water is maintained above about 7.5 because most of the arsenic(V) can be expected to be present either in the form of $HAsO_4{}^{2-}$ or $H_2AsO_4{}^-$. H_3AsO_3 remains in the neutral form at above about 7.5; therefore, As(III) in water is required to be converted into As(V) by a suitable oxidizing agent, such as chlorine. There is a risk of degradation of the resin by the oxidizing agents during the oxidation of the arsenic ($+3$) to achieve a ($+5$) oxidation state. For the effective removal of arsenic, the suitable conditions are that the arsenic has a ($5+$) oxidation state and that the pH be at least 7.5. Experimental works show that ion exchange can achieve arsenic reductions of more than 95%. The ion exchange process has the disadvantage of releasing noxious chemical reagents used in the resin regeneration into the environment.

The major disadvantages of the precipitation, adsorption, and ion exchange methods are the requirements of multiple chemical treatments, pre- and/or posttreatment of drinking water, disciplined/trained operation, high running/capital cost, and more importantly, regeneration of the medium and handling of arsenic-contaminated sludge. Disposal of the sludge will probably pose a problem in most cases.

While adsorption-based processes are often suitable for domestic water purification or at the most for a small community, it has the associated problem of frequent replacement as regeneration at such level is practically impossible. In the study region, a large number of activated alumina adsorbent-based community water filters installed earlier have turned defunct. Large-scale physico-chemical treatment plants could be very effective for supply of arsenic-free water to a large community but this often needs government-level initial investment and continuous operating costs. Membranes with high selectivity have the potential to produce totally arsenic-free water due to the small molecular weight (<150 Da) of most dissolved species of arsenic. Arsenic occurs mostly either as trivalent arsenite or as pentavalent arsenate in natural water and these are part of arsenic acid (H_3AsO_4) and arsenious acid (H_3AsO_3) systems, respectively, protonation of which depends on the pH of the aqueous system. At typical pH conditions of 6.5 to 8.0, As(V) remains as an anion and As(III) as a neutral molecule. Thus, membranes have the potential to remove arsenic from drinking water. In the next section we therefore discuss membrane-based processes for arsenic removal.

1.5.4 Membrane Filtration

Based on the main driving force, which is applied to accomplish the separation, many membrane processes can be distinguished.

1.5.4.1 Pressure-Driven Membrane Filtration

Pressure-driven membrane processes are commonly divided into four overlapping categories of increasing selectivity: microfiltration (MF), ultrafiltration (UF), nanofiltration (NF), and hyperfiltration or reverse osmosis (RO). MF is characterized by a membrane pore size between 0.05 and 2 μm and operating pressures below 2 bars. MF is used primarily to separate particles and bacteria from other smaller solutes. UF is characterized by a membrane pore size between 2 nm and 0.05 μm and operating pressures between 1 and 10 bars. UF is used to separate viruses, colloids like proteins from small molecules like sugars and salts. NF is characterized by a membrane pore size between 0.5 and 2 nm and operating pressures between 5 and 40 bars. NF is used to achieve a separation between sugars, other organic molecules, and multivalent salts on one hand and monovalent salts and water on the other. NF relies on physical rejection based on molecular size and charge. RO membrane contains extremely small pores (<0.001 μm).

Transport of the solvent is accomplished through the free volume between the segments of the polymer of which the membrane is constituted. The operating pressures in RO are generally between 10 and 100 bars and this technique is used mainly to remove water. NF is also known as a membrane softening process for its ability to remove the divalent ions in water that cause hardness (i.e., calcium and magnesium). NF also has the ability to remove sulfate as well as lesser quantities of the monovalent dissolved solids such as chloride and sodium. On the other hand, RO operates at higher pressure with greater rejection of all dissolved solids such as chloride and sodium. Separation is accomplished by MF membranes and UF membranes via mechanical sieving, while capillary flow or solution diffusion is responsible for separation in NF membranes and RO membranes. According to the pressure applied, pressure-driven membrane filtration is classified into two categories: low pressure membrane filtration, such as microfiltration (MF) and ultrafiltration (UF), and high pressure filtration, such as nanofiltration (NF) and reverse osmosis (RO).

1.5.5 Electrodialysis

Instead of applied pressure on the feed side in NF or RO, an electric field is applied in Electrodialysis (ED) to draw the ions of dissolved solids through the membranes leaving the fresh water behind. The cations (such as calcium and magnesium) are attracted to a negatively charged electrode and the anions (such as sulfate and arsenic) are attracted to a positively charged electrode. The membranes separating the electrodialysis unit are made up of cation and anion exchange resins. As electrodialysis is more effective in removing As(V) than As(III) like other arsenic removal processes, an oxidizing agent chlorine should be added to the feed for converting AS(III) to As(V). Oxidizing agents are harmful to conventional ion exchange resins (from which the membranes are made). And, as with reverse osmosis, electrodialysis tends to remove much more from the water than just the arsenic. The product water from an electrodialysis unit could be too good, possibly requiring posttreatment to meet municipal drinking water standards.

1.5.6 Temperature-Driven Membrane Filtration

Temperature-driven membrane filtration, known as membrane distillation (MD), is the emerging modern membrane separation process in which water vapor transports through a microporous hydrophobic membrane with pore sizes ranging from 0.1 to 1 μm. As the MD process allows only vapor through microporous membrane, 100% (theoretical) of arsenic along with other ions, macromolecules, colloids, cells, and other nonvolatile constituents present in arsenic-contaminated water are rejected. Pressure-driven processes such as RO, UF, pervaporation (PV), and MF have not been shown to achieve such high levels of rejection. Moreover, the MD process does not require any oxidant for conversion of As(III) to As(V).

1.5.7 Hybrid Methods of Arsenic Removal

Hybrid methods combine two or more conventional methods. As an example, chemical treatment may be combined with physical filtration like membrane separation. Hybrid methods overcome the limitations of the individual component methods and often result in much better separation in relatively less time. Membrane-integrated hybrid methods have been found to remove arsenic from contaminated groundwater with a very high degree of separation efficiency.

REFERENCES

[1] Morgada ME, Levy EK, Salomone V, Farias SS, Gerado L, Litter MI. Arsenic (V) removal with nanoparticulate zerovalent iron: effect of UV light and humic acids. Catal Today 2009;143(3–4):261–8.

[2] Heredia OS, Cirelli AF. Trace elements distribution in soil, pore water and groundwater in Buenos Aires, Argentina. Geoderma 2009;149(3–4):409–14.

[3] Chowdhury UK, Biswas BK, Chowdhury TR, Samanta G, Mandal BK, Basu GC, et al. Groundwater arsenic contamination in Bangladesh and West Bengal, India. Environ Health Perspect 2000;108(5):393–7.

[4] Chakraborti D, Mukherjee SC, Pati S, Sengupta MK, Rahman MM, Chowdhury UK, et al. Arsenic groundwater contamination of in Middle Ganga Plain, Bihar, India: a future danger? Environ Health Perspect 2003;111:1194–201.

[5] Harvey CF, Ashfaque KN, Yu W, Badruzzaman ABM, Ali MA, Oates PM, et al. Groundwater dynamics and arsenic contamination in Bangladesh. Chem Geol 2006;228:112–36.

[6] Klump S, Kipfer R, Cirpka O, Harvey CF, Brenwald M, Ashfaque K, et al. Groundwater dynamics and arsenic mobilization in Bangladesh assessed using noble gases and tritium. Environ Sci Technol 2006;40:243–50.

[7] Guha Mazumder DN. Chronic arsenic toxicity: clinical features, epidemiology, and treatment: experience in West Bengal. J Environ Sci Health A Tox Hazard Subst Environ Eng 2003;38:141–63.

[8] Pal P, Sen M, Manna AK, Pal J, Pal P, Roy SK, et al. Contamination of groundwater by arsenic: a review of occurrence, causes, impacts, remedies and membrane-based purifications. J Integr Environ Sci 2009;6(4):1–22.

[9] Das D, Chatterjee A, Samanta G, Mandal BK, Chowdhury TR, Chowdhury PP, et al. Arsenic contamination in groundwater in 6 districts of West Bengal, India - the biggest arsenic calamity in the world. Analyst 1998;119:168–N170.

[10] Bhattacharjee S, Chakravarty S, Maity SV, Dureja V, Gupta KK. Metal contents in the ground water of Sahebgunj district, Jharkhand, India, with special reference to arsenic. Chemosphere 2005;58:1203–17.

[11] Panthi SR, Sharma S, Mishra AK. Recent status of arsenic contamination in groundwater of the terai region of Nepal. J Sci Eng Tech 2006;1:1–11.

[12] Pokhrel D, Bhandari BS, Viraraghavan T. Arsenic contamination of groundwater in the Terai region of Nepal: an overview of health concerns and treatment options. Environ Int 2009;35(1):157–61.

[13] Welch AH, Westjohn DB, Helsel DR, Wanty RB. Arsenic in groundwater of the United States: occurrence and geochemistry. Groundwater 2000;38: 589–604.

[14] Nickson RT, McArthur JM, Shrestha B, Kyaw-Myint TO, Lowry D. Arsenic and other drinking water quality issues, Muzaffargarh District, Pakistan. Appl Geochem 2005;20(1):55–68.

[15] World Bank Technical Report, 2004. Arsenic occurrence in Groundwater in South and East Asia-Scale, Causes and Mitigation, Report No. 31303.

[16] Xia S, Dong B, Zhang Q, Xu B, Gao N, Causseranda C. Study of arsenic removal by nanofiltration and its application in China. Desalination 2007;204:374–9.

[17] Wickramasinghe SR, Han B, Zimbron J, Shen Z, Karim MN. Arsenic removal by coagulation and filtration: comparison of ground waters from the United States and Bangladesh. Desalination 2004;169:231–44.

[18] Tong NT. ESCAP – IWMI Seminar on Environmental and Public Health Risks Due to Contamination of Soils, Crops, Surface and Groundwater from Urban, Industrial and Natural Sources in South East Asia. Hanoi, Vietnam. 10th December – 12th December, 2002.

[19] Berg M, Tran HC, Nguyen TC, Pham HV, Schertenleib R, Giger W. Arsenic contamination of ground water and drinking water in Vietnam: a human health threat. Environ Sci Technol 2001;35:2621–6.

[20] Romero-Schmidt H, Naranjo-Pulido A, Méndez- Rodríguez L, Acosta-Vargas B, Ortega-Rubio A. Environmental health risks by arsenic consumption in water wells in the Cape region, Mexico. In: Brebbia CA, Fajzieva D, editors. Southhampton, UK: WIT Press; 2001.

[21] Smedley PL, Kinniburgh DG. A review of the source, behaviour and distribution of arsenic in natural waters. Appl Geochem 2001;17:517–68.

[22] Hsieh LC, Weng Y, Huang CP, Li KC. Removal of arsenic from groundwater by electro-ultrafiltration. Desalination 2008;234:402–8.

[23] Tseng CH. Abnormal current perception thresholds measured by neurometer among residents in Blackfoot disease-hyperendemic villages in Taiwan. Toxicol Lett 2003;146 (1):27–36.

[24] USEPA. WHO reports. Various reports available from US-Environmental Protection Agency website: www.epa.gov/safewater/arsenic.html and WHO, www.who.int/water_sanitation_health/water_quality/arsenic.htm; 2001 [accessed 15.11.08].

[25] Rahman MM, Mukherjee D, Sengupta MK, Chowdhury UK, Lodh D, Chanda CR, et al. Effectiveness and reliability of arsenic field testing kits: are the million dollar screening projects effective or not. Environ Sci Technol 2002;36(24):5385–94.

[26] Schnoor JL, editor. Environmental modeling. Fate and transport of pollutants in water, air, and soil. NY: John Wiley & Sons; 1996.

[27] Hung DQ, Nekrassova O, Comptom RG. Analytical method for inorganic arsenic in water: a review. Talanta 2004;64:269–77.

[28] MINEQL plus version 4.01. A chemical equilibrium modeling system. Environmental Research Software; 1998.

[29] Das D, Chatterjee A, Samanta G, Mandal BK, Samanta G, Chanda C, et al. Arsenic in ground water in six districts of West Bengal, India. The biggest arsenic calamity in the world. Part 2. Arsenic concentration in drinking water, hair, nail, urine, skin-scale and liver tissue (biopsy) of the affected people. Analyst 1995;120:917–24.

[30] Smith AH, Lingas EO, Raman M. Contamination of water by arsenic in Bangladesh: a public health emergency. Bull World Health Organ 2000;78:1093–2103.

[31] Akai J, Izumi K, Fukuhara H, Nakano HS, Yoshimura T, Ohfuji H, et al. Minerological and geomicrobiological investigations of groundwater arsenic enrichment in Bangladesh. Appl Geochem 2004;19:215–30.

[32] Islam FS, Gault AG, Boothman C, Polya DA, Charnock JM, Chatterjee D, et al. Role of metal-reducing bacteria in arsenic release from Bengal delta sediments. Nature 2004;430:68–71.

[33] Sengupta S, McArthur JM, Sarkar A, Leng MI, Ravenscroft P, Howarth RJ, et al. Do ponds cause arsenic-pollution of groundwater in Bengal Basin? An answer from West Bengal. Environ Sci Tech 2008;42:5156–64.

[34] Acharya SK. Arsenic contamination in groundwater affecting major parts of Southern West Bengal and parts of Western Chhattisgarh: source and mobilization process. Curr Sci 2002;82(6):740–3.

[35] Smedley PL, Nicolli HB, Macdonald DMJ, Barros AJ, Tullio JO. Hydro geochemistry of arsenic and other inorganic constituents in groundwater from La Pampa, Argentina. Appl Geochem 2002;17:259–84.

[36] Pokhrel D, Bhandari BS, Viraraghavan T. Arsenic contamination of groundwater in the Terai region of Nepal: an overview of health concerns and treatment options. Environ Int 2009;35(1):157–61.

[37] McArthur JM, Ravenscroft P, Safiulla S, Thirlwall MF. Arsenic in groundwater: testing pollution mechanisms for sedimentary aquifers in Bangladesh. Water Resour Res 2001;37:109–17.

[38] Welch AH, Westjohn DB, Helsel DR, Wanty RB. Arsenic in groundwater of the United States: occurrence and geochemistry. Groundwater 2000;38:589–604.

[39] Oremland RS, Newman, Kail B, Stolz J. Bacterial respiration of arsenate and its significance in the environment. In: Frankenberger WQ, editor. Environmental chemistry of arsenic. New York: Marcel Dekker; 2002 [chapter 11].

[40] Islam FS, Gault AG, Bootham C, Poyla DA, Charnock JM, Chatterjee D, et al. Role of metal-reducing bacteria in arsenic release from Bengal Delta Sediments. Nature 2004;430:68–71.

[41] Sanders JG. Arsenic cycling in marine systems. Mar Environ Res 1980;3:257–66.

[42] Cappuyns V, Herreweghe SV, Swennen R, Ottenburgs R, Deckers J. Sci Total Environ 2002;295:217–40.

[43] Tamaki S, Frankenberger WT. Environmental biochemistry of arsenic. Rev Environ Contam Toxicol 1992;124:79–110.

[44] Shibata Y, Morita M, Fuwa K. Selenium and arsenic in biology: their chemical forms and biological functions. Adv Biophys 1992;28:31–80.

[45] Wang S, Mulligan CN. Occurrence of arsenic contamination in Canada: sources, behaviour and distribution. Sci Total Environ 2006;366(2–3):701–21.

[46] Ming-Cheng S. An overview of arsenic removal by pressure-driven membrane processes. Desalination 2005;172:85–97.

[47] Mascher R, Lippmann B, Bergmann H. Arsenic toxicity: effects on oxidative stress response molecules and enzymes in red clover plants. Plant Sci 2002;163(5):961–9.

[48] Squibb KS, Fowler BA. The toxicity of arsenic and its compounds. In: Fowler BA, editor. Biological and environmental effects of arsenic. Amsterdam: Elsevier; 1983. p. 233–69.

[49] Dopp E, Hartmann LM, Florea AM, van Recklinghausen U, Pieper R, Shokouhi B, et al. Uptake of inorganic and organic derivates of arsenic associated with induced cytotoxic and genotoxic effects in Chinese hamster ovary (CHO) cell. Toxicol Appl Pharmacol 2004;201:156–65.

[50] Jain CK, Ali I. Arsenic: occurrence, toxicity and speciation techniques. Water Res 2000;34:4304–12.

[51] Ning RY. Arsenic removal by reverse osmosis. Desalination 2002;143:237–41.

[52] Roy M, Nilsson L, Pal P. Development of groundwater resources in a region with high population density: a study of environmental sustainability. Environ Sci 2008;5(4):251–67.

[53] Wickramasinghe SR, Han B, Zimbron J, Shen Z, Karim MN. Arsenic removal by coagulation and filtration: comparison of ground waters from the United States and Bangladesh. Desalination 2004;169:231–44.

[54] Majumdar DN. Report on health effects of arsenic toxicity. WB: CGWB; 2002, p. 81.

[55] Karim MDM. Arsenic in groundwater and health problems in Bangladesh. Water Res 2000;34:304–10.

[56] Tseng CH. Abnormal current perception thresholds measured by neurometer among residents in blackfoot disease-hyperendemic villages in Taiwan. Toxicol Lett 2003;146 (1):27–36.

[57] Chen Y, Parvez F, Gamble M, Islam T, Ahmed A, Argos M, et al. Arsenic exposure at low-to-moderate levels and skin lesions, arsenic metabolism, neurological functions, and biomarkers for respiratory and cardiovascular diseases: review of recent findings from the Health Effects of Arsenic Longitudinal Study (HEALS) in Bangladesh. Toxicol Appl Pharmacol 2009;239(2):184–92.

[58] Tsai SY, Chou HY, Chem CM, Chen CJ. The effects of chronic arsenic exposure from drinking water on the neurobehavioral development in adolescence. Neurotoxicology 2003;24(4–5):747–53.

[59] Wasserman GA, Liu X, Parvez F, Ahsan H, Factor-Litvak P, van Geen A, et al. Water arsenic exposure and children's intellectual function in Araihazar Bangladesh. Environ Health Perspect 112:1329–33.

[60] Mallick P, Chakrabarti Mallick J, Guha B, Khuda-Bukhsh AR. Ameliorating effect of micro doses of a potentized homeopathic drug, Arsenicum Album, on arsenic-induced toxicity in mice. BMC Complement Altern Med 2003;7. http://dx.doi.org/10.1186/1472-6882-3-7.

[61] Murphy RND. Homoeopathic medical repertory. 2nd revised ed. New Delhi: B. JAIN Publishers Ltd; 2006, Reprint.

[62] Niu S, Cao S, Shen E. The status of arsenic poisoning in China. In: Abernathy CO, Caledron RL, Chappel WR, editors. Arsenic: exposure and health effects. London: Chapman and Hill; 1997. p. 76–83.

[63] Patrick L. Toxic metals and antioxidants, Part II. The role of antioxidants in arsenic and cadmium toxicity. Alternative Med Rev 2003;8:106–27.

[64] Verret WJ, Chen Y, Parvez F, Alauddin M, Islam T, Graziano JH, et al. A double-blind placebo-controlled trial to evaluate the efficacy of vitamin E and selenium supplementation on arsenic-induced skin lesions in Bangladesh. J Occup Environ Med 2005;47 (10):1026–35.

[65] Braunwald F, Kasper H, Longo J. Harrison's principles of internal medicine. 15th international ed. New York: McGraw Hill; 2001.

CHAPTER 2

Chemical Treatment Methods in Arsenic Removal

Contents

Groundwater Arsenic Remediation
http://dx.doi.org/10.1016/B978-0-12-801281-9.00002-3

2.1 DIFFERENT FORMS OF ARSENIC IN GROUNDWATER

Arsenic can occur in the environment in various forms and oxidation states (-3, 0, $+3$, $+5$) but in natural water, arsenic occurs mainly in inorganic forms such as oxyanions of trivalent arsenite or as pentavalent arsenate. The two oxidation states common in drinking water in the form of arsenate and arsenite are part of the arsenic (H_3AsO_4) and arseneous (H_3AsO_3) acid systems, respectively. These two forms depend upon oxidation–reduction potential and pH of the water. At typical pH values of 5.0–8.0 in natural waters, the predominant arsenate species are $H_2AsO_4^-$ and $HAsO_4^{2-}$, and the arsenite species is H_3AsO_3. Under oxidizing conditions, $HAsO_4^{2-}$ dominates at a high pH regime, whereas H_3AsO_4 predominates at a low pH regime. $H_2AsO_4^-$ predominates at a low pH (<6.9). This means that As(III) remains as a neutral molecule in natural water. Arsenates are stable under aerobic or oxidizing conditions, while arsenites are stable under anaerobic or mildly reducing conditions. In reducing waters, arsenic is found primarily in the trivalent oxidation state in the form of arseneous acid, which ionizes according to the following equations:

$$H_3AsO_3 \rightleftharpoons H^+ + H_2AsO_3^- \qquad pK_a = 9.22$$

$$H_2AsO_3 \rightleftharpoons H^+ + HAsO_3^{2-} \qquad pK_a = 12.3$$

The acid base dissociation reactions of arsenic acid can be described as:

$$H_3AsO_4 \rightleftharpoons H^+ + H_2AsO_4^- \qquad pK_a = 2.20$$

$$H_2AsO_4 \rightleftharpoons H^+ + HAsO_4^{2-} \qquad pK_a = 6.97$$

$$HAsO_4^2 \rightleftharpoons H^+ + AsO_4^{3-} \qquad pK_a = 11.53$$

Surface water is also found to be contaminated with arsenic by the anthropogenic sources to various degrees since arsenic is also used in agriculture (pesticide), industrial applications, mining activities, and feed additives.

2.2 CHEMICAL PRECIPITATION

Arsenic can be separated from aqueous solutions through chemical precipitation, exploiting the insolubility of some arsenic compounds. Most dominant arsenic compounds that are precipitated out in this way are arsenic sulphide, ferric arsenate, and calcium arsenate, where pH plays a very crucial role in such precipitation. In the neutral pH regimes, the inorganic arsenic compounds of Cu(II), Zn(II), Pb(II), and Fe(II) are more stable [1]. In chemical precipitation, the As(V) is the dominant form. Iron (II) arsenate [2] is highly insoluble and stable for its successful adoption [2]. A large number of calcium arsenate compounds can be very effectively precipitated out from aqueous solutions of As(V) by raising pH through the addition of lime. But compounds such as those precipitated out at a pH above 8 are often not very stable, particularly in the atmospheric carbon dioxide environment where soluble carbonates are easily formed. More complex arsenic compounds such as apatite structured calcium phosphate arsenate or ferric arsenite have been found to be more appropriate forms of arsenic precipitation and subsequent stabilization.

Chemical precipitation in general is considered to be a permanent, efficient, and easy-to- monitor method that can have immediate results. Simultaneous removal of many metal contaminants is possible with the chemical precipitation method. Chemical precipitation may be very useful for large-scale treatment of high-arsenic water, but is not suitable for deep elimination of arsenic up to the level (10 ppb) prescribed by the World Health

Organization (WHO) for safe drinking water. Where chemical precipitation alone is not sufficient to meet the stringent regulations, this method may be used in conjunction with other methods. Such hybrid methods are discussed in Chapter 4. The major disadvantage of chemical coagulation-precipitation may be the generation and handling of huge quantities of sludge, and relatively high maintenance and operational costs. In chemical precipitation, calcium, aluminum, and ferric ions are widely used for precipitation-separation of arsenic from water.

2.2.1 Alum precipitation

In this coagulation–precipitation process, added aluminium ions reduce absolute values of zeta potential of the particles, resulting in coagulation–flocculation of the fine particles. Table 2.1 indicates the chemicals used in such treatment. Arsenic ions precipitate with aluminium ions being enmeshed in the coagulates–precipitates. Finally, separation of arsenic from water is effected through downstream sedimentation and filtration. The coagulation-precipitation process is pH-dependent. In the pH range of 5–7, the alum precipitation process is very effective. Coagulation–precipitation is found to remove As(V) more effectively than As(III). Thus for efficient removal of arsenic from water where both forms of arsenic are present or where only As(III) is present, it is necessary to convert As(III) in to As(V). Figure 2.1 indicates the effect of pH and chlorine on the arsenic removal process.

Table 2.1 Summary of Chemical Precipitation Processes in Arsenic Removal

Chemical precipitation process	Initial arsenic concentration used, $\mu g/L$	Chemicals used	pH	Arsenic removal %	
				Presence of chlorine	Absence of chlorine
Alum	300	Cl_2, acid, alum, NaOH	<6.5	90	20
Iron	300	Cl_2, $Fe_2(SO_4)_3$	6–8	90	60
Lime softening	400	Cl_2, lime, acid	>10.5	90	80
Coprecipitation	<100	Cl_2 needed Other chemicals $FeCl_3$, SO_2, $KMnO_4$	7+	40–90	–

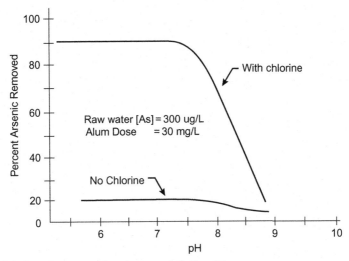

Figure 2.1 Arsenic removal by alum precipitation [6].

From Figure 2.1, it is clear that for the removal of arsenic, alum is most effective if the oxidizing agent chlorine is added ahead of the flocculator and clarifier and the pH is maintained at 7 or less.

2.2.2 Lime softening

The arsenic removal efficiencies of this process highly depend on the pH and the presence (or absence) of chlorine (shown in Figure 2.2).

Figure 2.2 indicates that the arsenic removal efficiency in the absence of chlorine increases steeply from 15% to 80% as the pH changes from 9 to 11 when the arsenic concentration of the contaminated water is 400 µg/L. Arsenic removal efficiency for the same feed varies almost linearly (from 30% to 95%) with pH in the range of 8–11 in the presence of chlorine. The produced sludge containing arsenic has no added value and can limit the use of technology. For this reason, treatment in two stages is justified: lime softening followed by arsenic removal.

2.2.3 Iron precipitation

In the iron precipitation process, ferric ions are added to arsenic-bearing water where arsenic coprecipitates with ferric hydroxides on being enmeshed in the coagulate–precipitate. The forms of the coagulants used in such treatment are presented in Table 2.1. A pH range of 5–8 is found suitable for removal of arsenic with ferric ions. Under the same conditions,

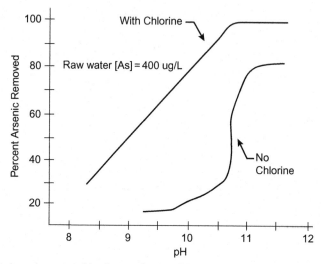

Figure 2.2 Arsenic removal by lime softening [6].

arsenic removal is reduced to about 50% or more in the absence of chlorine, as shown in Figure 2.3.

The pH adjustment does not appear to be as important as it does with the alum precipitation process. This process is well known for its simplicity, versatility, selectivity, and low cost.

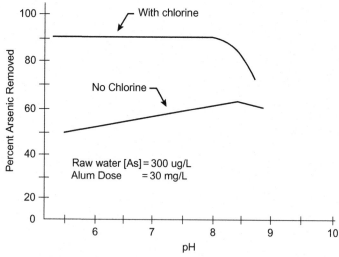

Figure 2.3 Arsenic removal by iron precipitation [6].

2.2.4 Enhanced coagulation

Arsenic separation by chemical precipitation can be done through enhanced coagulation by adding coarse particles like calcite. Smaller calcite particles (30–45 µm) can be more effective than larger ones by virtue of larger specific surface area. Arsenic-borne coagulates get coated on the surface of such calcite particles for eventual removal as precipitates. Arsenic-borne coagulates, coated onto the surface of the calcite particles, get bigger and more dense, facilitating easy and quick settling in the downstream sedimentation and filtration.

2.2.5 Coprecipitation

Two processes are employed for the removal of arsenic combined with iron (and/or manganese). One of the processes involves a proprietary media. In one variation of this process, chlorine is injected into raw water containing iron and/or manganese in a reaction vessel for 1–2 min. Sulfur dioxide may also be injected into water and allowed to react with iron and/or manganese for a short period of time. Water is then discharged into one or more filter vessels that contain the proprietary media. Arsenic reductions of perhaps 50% can be obtained with this process [3]. In another process, three additional chemicals—ferric chloride, potassium permanganate, polymeric aluminum silicate sulfate—are needed. In addition to the two reaction vessels and filters described in the first process, a flocculator and clarifier are also used. Iron chloride ($FeCl_3$) is added along with chlorine in the first reaction vessel. Following the first reaction vessel, a coagulant aid (PASS; polymeric aluminum silicate sulfate) is added in addition to the SO_2. The water then passes through the second reaction vessel. It is discharged from the second reaction vessel and the water flows into a flocculator and clarifier before being filtered. Arsenic removal rates of more than 90% may be obtained in the second process. Table 2.1 summarizes the precipitation processes.

2.3 PHYSICAL SEPARATION

Through chemical precipitation arsenic is mobilized from an aqueous phase to a solid phase. But this does not automatically ensure its separation from water, particularly when the precipitation is in the form of fine colloidal particles (1–100 µm). Turbidity of water reflects the presence of suspended particles and is measured through the Tyndall effect in a turbidity meter, which is reflection of white light (average wavelength 400–640 nm) by colloidal particles while passing through such suspensions.

A colloidal suspension is basically a stable phase showing little tendency to aggregate and separate from the aqueous phase. For separation of chemical precipitates effectively from the water phase, destabilization of the colloidal state is necessary. Because colloidal particles have a large surface-to-mass ratio, the behavior of the colloidal suspension basically represents surface phenomena. Colloidal particles assume electrical charges with respect to the surrounding environment. This is observed in the migration of such charged particles toward the pole of the opposite charge when these are placed in an electrical field. This phenomenon is known as electrophoresis. Thus destabilization of colloids means neutralization of such a surface charge. The effective way of destabilization is to aggregate the fine particles through coagulation. Driving particles together for destabilization is called coagulation. Stability of colloids is explained most explicitly by the diffuse–double–layer theory.

2.3.1 Diffuse-double-layer theory

Diffuse-double-layer theory states that as colloidal particles assume a positive or negative charge due to the presence of charged groups within, or adsorption of a charged layer from, the surrounding medium, an electrical double layer of the opposite charge is formed at the interface between the solid phase and the aqueous phase to ensure electroneutrality of the overall colloidal system. A fixed covering of positive ions is formed over a group of negatively charged particles. This fixed layer of charge is called the Stern layer, which in turn is surrounded by a thin movable layer of positive charges called the diffuse layer, as shown in Figure 2.4.

Counter ions of the aqueous phase are electrostatically attracted to the colloid surface with an opposite charge. Concentration of the counter ions is naturally high in the immediate vicinity of the colloidal particle, and the same diffuse out as the distance between the surface of the colloidal particle and the bulk solution increases. The magnitude of the charge at the surface of shear is called the zeta potential, which can be measured through electrophoresis.

The zeta potential is expressed as

$$\psi = 4\pi\delta q/D$$

where q is the charge of the particle, δ is the thickness of the zone of influence of the charge, and D is the dielectric constant of the medium.

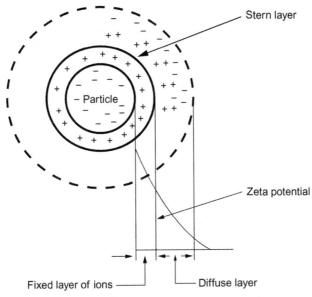

Figure 2.4 Diffuse-double-layer diagram.

Between two similarly charged colloidal particles, both a repulsive force as well as an attractive van der Waal's force work. If these repulsive and attractive forces are designated as V_R and V_A, respectively, then the net interactive force or energy may be expressed as $V_R - V_A$. This is called an energy barrier of double-layer interaction. The stability of the colloidal suspension basically depends on this net energy barrier. For destabilization, sufficient kinetic energy needs to be added to the colloidal system to overcome this energy barrier. The mathematical expression of the potential repulsive energy of electrical double-layer interaction in a suspension of heterogeneous particles between two heterogeneous spheres may be written as [4]:

$$V_R = 32\pi\varepsilon_0\varepsilon_r \frac{2a_1a_2}{(a_1 + a_2)} \left(\frac{kT}{ev}\right)^2 \tanh\left(\frac{ev\psi_1}{4kT}\right) \tanh\left(\frac{ev\psi_2}{4kT}\right) \exp(-kh) \quad (2.1)$$

where a_1 and a_2 are the radii of the particles 1 and 2, respectively. ε_0 stands for the vacuum dielectric permittivity and ε_r for the relative dielectric permittivity of the medium. k is the Boltzmann constant and T is the absolute temperature. Elementary charge and ionic valence of the electrolyte are designated as e and v, respectively. ψ_1 and ψ_2 are the outer Helmholtz plane (OHP) potentials or zeta potentials of particles 1 and 2, respectively.

k is the Debye reciprocal length where h is the shortest separation between the two particles. The potential energy (V_A) of van der Waals interaction between two heterogeneous particles is expressed as [5]:

$$V_A = -\frac{a_1 a_2}{(a_1 + a_2)} \left(\frac{A_{132}}{6h} \right)$$ (2.2)

where A_{132} is the Hamaker constant of particles 1 and 2 in medium 3, which may be obtained as [6]:

$$A_{132} = \left(\sqrt{A_{11}} - \sqrt{A_{33}} \right) \left(\sqrt{A_{22}} - \sqrt{A_{33}} \right)$$ (2.3)

where A_{11}, A_{22}, and A_{33} are the Hamaker constants of particles 1, 2, and medium 3 in vacuum.

If A_{11} represents the Hamaker constant of the arsenic-borne coagulates and A_{22} stands for the Hamaker constant of precipitation aid like $Fe(OH)_3$, and that for water is represented by A_{33}, then $A_{11} > A_{33}$ and $A_{22} > A_{33}$. Thus $A_{123} > 0$, and $V_A < 0$ according to Eqs. (2.2) and (2.3), indicating an attractive interaction of van der Waals between the two heterogeneous particles. In the pH range of around 5–9, $\Psi_1 < 0$ (arsenic-borne coagulates), and $\Psi_2 > 0$ (ferric hydroxide), leading to $\tanh \left(\frac{ev\psi_1}{4kT} \right) < 0$ and $\tanh \left(\frac{ev\psi_2}{4kT} \right) > 0$, and thus $V_R < 0$ according to Eq. (2.1), which implies that the electrical double-layer interaction is attractive between the two heterogeneous particles here. Therefore, in the said pH range, the total potential energy of interaction between the arsenic-borne coagulate and the ferric hydroxide particle is attractive at every distance. This means that there is no potential energy barrier between the two particles (arsenic-bearing fine coagulates and ferric hydroxide or coarse calcite particles) that prevents them from coming together. This is the mechanism that helps arsenic-borne fine coagulates to get coated onto the surface of the precipitation aid ferric hydroxide or coarse calcite particles. This coprecipitation is also called enmeshment precipitation, where the settling rate turns much faster.

2.3.2 Destabilization of colloids and settling of particles

There are four basic mechanisms of destabilization of colloidal systems.
- Double layer compression
- Adsorption and neutralization of charge
- Enmeshment-precipitation
- Interparticle bridging

2.3.2.1 Double-layer compression

In this mechanism, normally coagulant ions in high concentration are added to the system when similarly charged ions of the coagulants are repelled but the oppositely charged ions are attracted by the primary charge of the colloidal particles, causing compression of the diffuse double layer. Thus the decrease of the diffuse double layer with high concentration of counter ions of the coagulants helps overcome the energy barrier of a colloidal system, thus destabilizing it. Coagulation increases with charge of the coagulant ion ($Al^{3+} > Ca^{2+}$).

2.3.2.2 Adsorption and neutralization of charge

In this case, the charge of colloidal particles is neutralized by the addition of molecules of opposite charge that adsorb onto the surface of the colloidal particles. Overdosing such charged molecules may lead to restabilization of the system by the residual charges of the added molecules after neutralization of the primary charges of the colloidal particles.

2.3.2.3 Enmeshment-precipitation

In this case, added metal salts such as aluminum sulfate, ferric chlorides, calcium oxides, and so on precipitate as hydroxides, in which the colloidal particles get enmeshed and coprecipitate.

2.3.2.4 Interparticle bridging

Lamer (1963) proposed this mechanism where some long chain charged polymeric molecules are added to a colloidal system. One charged end of the polymer molecule attaches to a site of the colloid and the other end extends to the bulk solution. If the other end attaches to another colloidal particle then an effective bridging between two colloidal particles takes place, resulting in their settling together.

2.3.3 Filtration

2.3.3.1 Rapid sand filtration

The precipitated arsenic-borne coagulates get largely separated from water in the downstream settling or clarification units following the standard principles of settling. After separation of the large flocks, overflow from settling basins passes through filter beds for separation of the finer suspensions. For large-scale filtration, sand or silica, being abundantly available, is very widely used in large sand beds. Such sand filters are used in rapid mode where water passes to a clear well through a bed of sand filter. Only sand with a specified grain size is used. Thus sand is first properly sieved and is used as per the desired criteria of effective size and uniformity coefficient. If D_{10} and D_{60} stand for sieve sizes that do not allow more than 10% and 60% by weight of sand, respectively, to pass through then the effective size is taken as D_{10} and the uniformity coefficient is expressed by D_{60}/D_{10}. If P_{10} and P_{60} stand for the percentage of stock sand smaller than D_{10} and D_{60}, respectively, then sand with a grain size between D_{10} and D_{60} forms half of the specified sand. In other words, the usable stock sand (by percentage) may be expressed as:

$$P_{usable}/2 = (P_{60} - P_{10}) \qquad (2.4)$$

During filtration, the major problem often encountered is air binding, resulting from the release of dissolved air or gases in water. Such released air makes its way through the sand pores, effectively preventing water flow and resulting in rapid head loss. To overcome the problem of head loss, air-saturated water should be avoided and care should be taken to prevent algal growth.

2.3.3.2 Backwashing

The filter bed is cleaned by occasional backwashing. A water jet is used in the reverse direction, causing the entire sand bed to be fluidized. Occasionally the problem of mud ball formation is encountered during backwashing. Mud balls are formed by a coating of mud on sand grains when the mud layer from the top of the bed makes its way down through the sand pores. Mud balls sink further down because of their increased weight, preventing effective backwashing. This can be prevented by breaking the balls, replacing the sand, or adding 2–5% caustic soda solutions.

2.4 MODELING AND SIMULATION OF THE PHYSICO-CHEMICAL PROCESSES FOR SCALEUP

2.4.1 Introduction

For large-scale treatment of arsenic-contaminated groundwater in the arsenic-affected areas of the developing countries, there is hardly any better alternative to physico-chemical treatment like coagulation–precipitation. It bears a low treatment cost for a reasonably high degree of purification of a huge volume of contaminated water. However, if a high degree of separation is required following WHO guidelines, then such treatment may not be able to stringently meet the requirement. Where the alternate surface water source is far from the arsenic-affected villages, this low-cost physico-chemical treatment is likely to be the most promising one.

Physico-chemical separation through chemical coagulation and precipitation has been demonstrated as one of the most effective methods of arsenic separation [7, 8]. A number of combinations of coagulants and oxidants have been found to be effective under optimum pH conditions [9, 10]. It has been found that under optimum conditions of coagulation (by ferric chloride or alum), oxidation (by chlorine or potassium permanganate), and sedimentation and filtration, 95–98% arsenic removal is possible. It has been established that soluble arsenic can be effectively removed from drinking water by ferric chloride coagulation with prior oxidation of trivalent arsenic to pentavalent state using $KMnO_4$ or chlorine as an oxidant. Sedimentation followed by sand filtration can yield the results in reasonable time and at a reasonable cost. And in this treatment, pH has to be properly adjusted and maintained so as to take the feed water to the point of maximum insolubility of arsenic. Though either $KMnO_4$ or chlorine has been suggested as a possible oxidizing agent, use of chlorine should be discouraged in view of the possibility of the formation of carcinogenic chlorine by-products from the reactions of naturally occurring organic matter (NOM) with chlorine. Thus the most appropriate arsenic removal scheme should include in sequence one oxidation unit (with only $KMnO_4$ as oxidant), a coagulator (with $FeCl_3$ as coagulant and provision for controlling pH), a flocculator and a sedimentation unit, followed by one sand filtration unit. Such a treatment scheme is presented in Figure 2.5.

2.4.2 Operation of the treatment plant

The physico-chemical treatment system consists of one stirred oxidation reactor, a high-mixing coagulator, a slowly agitated flocculator, a sedimentation unit, followed by a sand filtration unit. The cylindrical oxidizer unit is provided with a mechanical agitator with three impellers. Baffles are fitted to the reactor unit. A slow agitator is provided in the flocculator unit.

The sedimentation unit follows the flocculator unit, and a sand filtration unit follows the sedimentation unit. The filter medium is granular sand having an average diameter of 0.00001 m. The set-up is run in continuous mode. $KMnO_4$ solution prepared in deionized water is added in the oxidation unit instantly with the introduction of the feed solution. Ferric chloride solution (prepared in distilled water for coagulation) is added to the coagulator in the same fashion. In groundwater (anaerobic condition), arsenic remains mainly in trivalent form. For oxidation of As(III) to As(V),

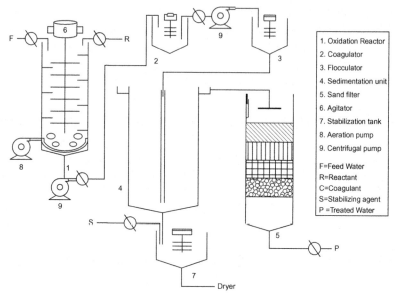

F=Feed Water
R=Reactant
C=Coagulant
S=Stabilizing agent
P =Treated Water

1. Oxidation Reactor
2. Coagulator
3. Flocculator
4. Sedimentation unit
5. Sand filter
6. Agitator
7. Stabilization tank
8. Aeration pump
9. Centrifugal pump

Figure 2.5 Typical physico-chemical treatment plant.

0.5–40.0 ppm KMnO$_4$ is added to the oxidation unit that is provided with a mechanical agitator. Because oxidation of arsenic or iron is favored at a low pH, a pH of 5.5 is maintained in the oxidation unit. The oxidation unit followed a coagulator unit where 0.5–200 ppm FeCl$_3$ is added as coagulant and caustic soda (NaOH) is added to adjust the pH to 7.6. The aqueous stream from the coagulator is then made to pass through the flocculator unit, which is provided with a low-speed agitator that facilitates particle-to-particle contact for larger flock formation and subsequent settling. A downstream sedimentation unit receives the stream from the flocculator unit for separation of the precipitates from the aqueous phase through sedimentation. The suspended solids that fail to settle in the sedimentation unit get separated from the stream while passing through a sand filter bed. Table 2.2 presents a set of typical operating parameters.

Samples from the outlet of the filtration unit may be analyzed in an atomic absorption spectrophotometer through the flame-FIAS technique. In this flame-FIAS technique, oxy-acetylene flame is used to atomize the sample element and FIAS (Flow Injection Analysis System) is used to inject an exact, reproducible volume of sample into a continuously flowing carrier system. The FIAS includes a peristaltic pump module, a flow injection valve, chemifold, gas/liquid separator, and flexible silicone rubber tubing.

Table 2.2 Typical Operational Conditions and Model Parameters

Temperature maintained in the units	302–305 K
Root mean square velocity gradient in the coagulator (G_1)	800 s^{-1}
Root mean square velocity gradient in the flocculator (G_2)	70 s^{-1}
Feed water flow rate	1.32 dm^3/min
Arsenic concentration of the feed water	1–2 mg/L
Oxidation rate constant (k)	3.23×10^{-3} s^{-1}
pH in the oxidation unit	6.5
pH in the coagulator and flocculator	7.6–8.0
Overall settling rate constant (k_{QM})	1.93×10^{-3} mol^{-1}s^{-1}
Coagulant concentration	30 mg/L
Oxidant concentration	15 mg/L

2.4.3 Measuring arsenic concentration in water

Analysis of the samples for arsenic concentration may be done following the flame-FIAS technique in an atomic absorption spectrophotometer. In the flame-FIAS technique, arsenic is analyzed after its conversion to volatile hydride, which was formed through the following reactions:

$$NaBH_4 + 3H_2O + HCl = H_3BO_3 + NaCl + 8H \qquad (2.5)$$

$$As^{3+} + H^{\cdot}(\text{excess}) \longrightarrow AsH_3 + H_2(\text{excess}) \qquad (2.6)$$

Volatile hydride is transported to the quartz cell of the atomic absorption spectrophotometer where it is converted to gaseous arsenic metal atoms at 1,173 K in air-acetylene flame; analysis should be done at 193.7 nm wavelength using a hollow cathode lamp. Samples to be analyzed for arsenic is prereduced (As^{5+} to As^{3+}) using a reducing solution containing 5% (w/v) potassium iodide (KI) and 5% (w/v) ascorbic acid. The reduced sample is allowed to stand at room temperature for 30 min. In hydride generation, 0.2% sodium borohydrate ($NaBH_4$) in 0.05% NaOH is used; 10% (w/v) HCl is used as a carrier solution. The hydride generator is purged using 99.99% pure nitrogen.

2.4.4 Computation of percentage removal of arsenic

The percentage removal of arsenic is computed using the initial value (C_{As0}) and the residual value (C_{As}) of arsenic concentration in feed water and treated water, respectively.

$$\% \text{ removal of arsenic} = (1 - C_{As0}/C_{As}) \times 100 \qquad (2.7)$$

2.4.5 Modeling and simulation of physico-chemical treatment process

Based on the treatment scheme as presented in Figure 2.5, a dynamic mathematical model is developed considering the involved process of kinetics and the hydrodynamics of the unit operations as described in the following.

2.4.5.1 Process kinetics and modeling basis

Arsenic generally occurs in inorganic form and in two valence states—As(III) and As(V). While As(V) species dominate under aerobic or oxidizing conditions, As(III) species dominate under reducing conditions. As(III) species may be present as arseneous acid (H_3AsO_3) and arsenite ions ($H_2AsO_3^-$, $HAsO_3^{2-}$, AsO_3^{3-}). As(V) exists as arsenic acid and arsenate ions. Effectiveness of arsenic separation depends on the physical and chemical characteristics of the arsenic species in water, particularly the valence state. As(V) precipitates more easily [7] than As(III); therefore, effects of the arsenic oxidation state can be eliminated by preoxidation of As(III) to As(V). In the presence of oxidizing agents like potassium permanganate ($KMnO_4$), chlorine (Cl_2), or hydrogen peroxide (H_2O_2), As(III) gets oxidized to As(V). This oxidation is very fast, and follows first-order kinetics as in the case of the oxidation of Fe^{2+} to Fe^{3+}.

This oxidation rate equation using $KMnO_4$ as an oxidant may be expressed as [11]:

$$\frac{d}{dt}\left[As^{III}\right] = -k\left[As^{III}\right] \tag{2.8}$$

where k is the first-order rate constant (s^{-1}).

Oxidation of iron and arsenic in the presence of the oxidizing agent $KMnO_4$ takes place following reactions (2.11) and (2.12), respectively, as shown:

$$KMnO_4 \longrightarrow K^+ + MnO_4^- \tag{2.9}$$

$$H_2O \longrightarrow H^+ + OH^- \tag{2.10}$$

$$3Fe^{2+} + MnO_4^- + 4H^- \longrightarrow 3Fe^{3+} + MnO_2 + 2H_2O \tag{2.11}$$

$$As^{3+} + MnO_4^- + 4H^+ \longrightarrow As^{5+} + MnO_2 + 2H_2O \tag{2.12}$$

Oxidation of As^{3+} to As^{5+} and that of Fe^{2+} to Fe^{3+} strongly depends on the concentration of $KMnO_4$ up to a certain level. In the oxidation reactor, a quick dispersion of the reaction ingredients is enhanced through rapid mixing over a short period. The oxidation reactor may, therefore, be assumed to be of CSTR type.

2.4.5.2 Modeling the process

Physico-chemical separation of arsenic from the aqueous phase to the solid phase following coagulation–flocculation–precipitation is basically a broad five-step process as described here.

1. Rapid mixing of coagulants like $FeCl_3$ or Alum ($Al_2(SO_4)_3.18H_2O$) takes place in the aqueous phase that contains arsenic. $FeCl_3$ has been found to be more effective than other coagulants in arsenic separation.
2. Nucleation of $Fe(OH)_3$ crystals takes place in the second step very quickly when $FeCl_3$ is added to the arsenic-containing aqueous solution.
3. Growth of the crystal particles takes place following the principles of orthokinetic flocculation due to both temporal and spatial variation of fluid velocity within the flocculator. Rate of change of concentration of the settling particles follows O'Melia. In the process of flocculation, a large number of small particles get converted into a small number of large particles.
4. Adsorption and enmeshment of arsenic onto the growing $Fe(OH)_3$ particles.
5. Coprecipitation of arsenic with metal hydroxides following different association mechanisms is the final step in the process when arsenic mainly coprecipitates as $As(V)–Fe(OH)_3$ flocks.

Though arsenic may be present in both trivalent as well as pentavalent form, coprecipitation of arsenic takes place from the aqueous solution as $As(V)–Fe(OH)_3$ following preoxidation of all trivalent arsenic into pentavalent form and subsequent adsorption onto ferric hydroxides since arsenic settles better in pentavalent form than in trivalent form. The following assumptions are involved in developing the model.

- Oxidation of trivalent arsenic into pentavalent form in presence of potassium permanganate follows a pseudo first-order reaction [11].
- Because of the quick mixing and dispersion requirements in the oxidation unit, the oxidation reactor may be assumed to be a CSTR type reactor.
- Because of spatial as well as temporal variation of the fluid velocity in the system, flocculation of arsenic precipitates may be assumed to follow an orthokinetic mechanism.
- For orthokinetic flocculation mechanism, change of concentration of settling particles may be assumed to follow O'Melia [12].
- The overall process of enmeshment of arsenic onto ferric hydroxides and subsequent settling may be assumed to follow a first order reaction kinetics in the backdrop of kinetic limitations and the difficulties in decoupling the interrelated phenomena.

The association mechanisms are inclusion, adsorption, occlusion, and solid solution formation. The inclusion mechanism is a mechanical entrapment of a portion of arsenic-containing solution within the growing crystal mass of $Fe(OH)_3$ whereas the adsorption mechanism involves attachment of arsenic onto the surface of crystal mass. In the occlusion mechanism, arsenic gets adsorbed onto the surface of growing crystal followed by further growth of crystal over the adsorbed arsenic mass. In the solid-solution mechanism, the solute of interest forms a solid solution with identical ions and coprecipitates as equal partners.

Thus removal of arsenic from the aqueous phase depends on a number of interdependent phenomena like formation of precipitate, coprecipitate and mixed precipitate, adsorption and enmeshment of inorganic arsenic species onto metal hydroxides, and subsequent settling through flocculation. It is extremely difficult to experimentally uncouple the effect of one phenomenon from that of the other. The overall process of formation of amorphous solid metal hydroxides with enmeshment of arsenic and its subsequent settling may be assumed to follow a second-order kinetics where rate of change of arsenic concentration depends on concentration of arsenic in the aqueous phase and concentration of coagulant. Separation of solid-phase arsenic takes place in the sedimentation and filtration units.

For a continuous treatment plant, settling is hardly complete in the sedimentation unit, and to separate the still-suspended particles, a filter such as a sand filter can be a very low cost and effective means for final separation of arsenic from drinking water. A dynamic mathematical model based on the assumptions discussed here is developed in the next section through mass balance in the units involved.

2.4.5.3 Material balance for the oxidizer unit

Overall mass balance of aqueous solution in the reactor unit:

$$\text{Change in mass} = \text{mass of raw water}\,|_{\text{input}} + \text{mass of oxidant}\,|_{\text{input}}$$
$$- \text{mass of treated water}\,|_{\text{output}}$$

$$\rho_0 A\left(\frac{dh}{dt}\right) = F_i \rho_i + F_{ri}\rho_{ri} - F_0\rho_0 \tag{2.13}$$

where ρ_i and ρ_0 are densities (kg/ m^3) of water at the inlet and outlet and ρ_r is the density of the oxidant. F_i and F_0 are volumetric flow rates (m^3/s) of the feed and treated water, respectively. F_r is volumetric feed rate (stoichiometric) of the oxidant. A is the reactor cross-sectional area and h is the liquid level in the reactor.

2.4.5.4 Component mass balance of arsenic

$$\text{Change in As(V) concentration} = \text{As(III) concentration}|_{\text{input}}$$
$$- \text{As(V) concentration}|_{\text{output}}$$
$$+ \text{generation of As(V)}$$

$$\frac{d}{dt}(C_A V) = F_i C_{A_i} - F_o C_A + V k C_A^{n_1} C_r^{n_2} \tag{2.14}$$

where C_{A_i} and C_A are the concentration $(kmol/m^3)$ of As(III) at the inlet and As(V) at the outlet of the reactor. C_r is the oxidant concentration $(kmol/m^3)$. k is the second-order reaction (oxidation) rate constant $(mol^{-1} s^{-1})$. n_1 is the kinetic constant and V is the volume of the reactor (m^3).

2.4.5.5 Component mass balance of oxidant

$$\text{Change in oxidant concentration} = \text{oxidant concentration}|_{\text{input}}$$
$$- \text{accumulation of oxidant}$$

$$\frac{d}{dt}(C_r V_r) = F_{r_i} C_{r_i} - V k C_A^{n_1} C_r^{n_2} \tag{2.15}$$

where C_{r_i} and C_r are the initial and instantaneous concentration $(kmol/m^3)$ of the oxidant and n_2 is kinetic constant.

2.4.5.6 Material balance of the coagulator and flocculator

Overall mass balance of the aqueous solution in the coagulator and flocculator units:

$$\text{Change in mass in the coagulator} - \text{flocculator} = \text{mass of the input stream}$$
$$+ \text{mass of the coagulant stream} - \text{mass of the output stream}$$

$$\rho_{QM_o} A_{QM} \left(\frac{dh_{QM}}{dt} \right) = F_{QM_i} \rho_{QM_i} + F_{c_i} \rho_{c_i} - F_{QM_o} \rho_{QM_o} \tag{2.16}$$

where ρ_{QM_i} and ρ_{QM_o} are the densities (kg/m^3) of the inlet and outlet aqueous solutions in the coagulator-flocculator and ρ_{C_i} is the density of the coagulant (kg/m^3). A_{QM} is the area of the coagulator/flocculator (m^2). $F_{QM_i}, F_{QM_o}, F_{C_i}$ are the flow rates (m^3/s) of the feed water, treated water, and coagulant.

Component mass balance of As(V):

Change in concentration of As(V) = As(V) concentration $|_{\text{input}}$

$\qquad\qquad$ − arsenic concentration $|_{\text{output}}$

$\qquad\qquad$ − accumulation of As(V)

$$\frac{d}{dt}\left(C_{QM_A}V_{QM}\right) = F_{QM_i}C_{QM_{A_i}} - F_{QM_0}C_{QM_A} - V_{QM}k_{QM}C_{QM_A}^{m_1}C_c^{m_2} \quad (2.17)$$

where $C_{QM_{A_i}}$ and C_{QM_A} are the concentrations (kmol/m^3) of arsenic at the inlet and outlet and C_c is the coagulant concentration (kmol/m^3) in the coagulator. V_{QM} is the volume of the coagulator (m^3). k_{QM} is the assumed overall second-order rate constant (mol^{-1} − s^{-1}) of arsenic flocculation, adsorption, enmeshment, and settling. m_1 and m_2 are the reaction kinetic constants.

Component mass balance of floc:

Change in floc concentration = generation of floc in the outlet stream

$$\frac{d}{dt}\left(C_{QM_{\text{floc}}}V_{QM}\right) = V_{QM}K_{QM}C_{QM_A}^{m_1}C_C^{m_2} - F_{QM_0}C_{QM_{\text{floc}}} \quad (2.18)$$

where $C_{QM_{\text{floc}}}$ is the concentration of the floc (kmol/m^3).

Component mass balance of coagulant:

Change in coagulant concentration = input concentration of coagulant

$\qquad\qquad\qquad\qquad$ − accumulation of coagulant

$$\frac{d}{dt}\left(C_cV_{QM}\right) = F_{C_i}C_{C_i} - V_{QM}k_{QM}C_{QM_A}^{m_1}C_c^{m_2} \quad (2.19)$$

Total rate of fall of floc concentration:

$$\frac{dC_{QM_{\text{floc}}}}{dt} = -\frac{2}{3}E_1\,G_1\,D_{QM_f}^3\,C_{QM_{\text{floc}}}^2 \quad (2.20)$$

$$E_1 = \frac{E'}{6.023 \times 10^{23}} \text{ and } E' = 6.023 \times 10^{23}$$

where $E_1 = 1(\text{mol}^{-1})$, $G_1(\text{s}^{-1})$ is the average root mean square velocity gradient in the coagulator-flocculator. D_{QM_f} (m) is the average diameter of the floc particles in the coagulator–flocculator.

2.4.5.7 Material balance for the sedimentation unit

$$\frac{dz}{dt} = \frac{G}{C_u} - U \quad (2.21)$$

where

$$G = F_0 \frac{C}{A_d}$$

$$C_u = C_{QM_{floc}}$$

F_0 is the volumetric feed rate (m³/s) of aqueous solution in the sedimentation unit. C is the floc concentration of the solution (kmol/m³). A_d is the sedimentation unit area (m²). C_u is the sludge concentration or floc concentration (kmol/m³) and U is the average settling velocity of the floc particles (m/s). dz/dt is the sedimentation rate (m/s) and G having unit kmol/m²s.

2.4.5.8 Filtration Unit
Filtrate flow rate:

$$\frac{dV_F}{dt} = \left[\frac{\mu \alpha W V_F}{A^2(-\Delta P)} + \frac{\mu R_m}{A(-\Delta P)} \right]^{-1} \tag{2.22}$$

where V_F is the volume of the filtrate (m³) and A is the area of the filter bed. α is the specific cake resistance (m/kg), and W is the solid concentration of the water to be filtered. μ is the viscosity of the aqueous solution at the inlet of filter unit. R_m is the filter medium resistance. $(-\Delta P)$ is the pressure drop through the filter medium and filter cake (N/m²).

The initial conditions are the following: $h=0$, $C_A=C_{A_i}$, $C_r=C_{r_i}$, $h_{QM}=0$, $C_{QM}=C_{QM_i}$, $C_{QM_{floc}}=0$, $C_C=C_{C_i}$, $h_{SM}=0$, $C_{SM_{floc}}=0$, $z=0$, and $V_F=0$.

Numerical solutions of the model equations are obtained using the modified Runge–Kutta–Gill method for the differential equations. A Visual Basic software program (ARSEPPA) was developed using these model equations. The software allows adjustment of integration step size so as to achieve a desired level of accuracy. In the present computation, the step size was adjusted to keep the relative error within 1%.

2.4.6 Determination of the model parameters
2.4.6.1 Computation of flow rate and concentration of oxidant
The flow rate of the oxidant was determined using a factor considering the stoichiometry of the reaction.

For flow and stoichiometric feed rate of oxidant dose:

$$F_{r_i} = f_1 F_i, \text{ where } f_1 < 1$$

$$C_{r_i} = \frac{F_i E_{As} C_{Ai} M_r}{M_{As} F_{r_i} E_r} \qquad (2.23)$$

where $E_{As}{:}E_r = 1$ mg:15 mg; and M_{As} and M_r are molecular weights of arsenic and oxidant, respectively.

The density of the treated water at the outlet was determined considering the average density of the feed raw water and the oxidant. It may be safely assumed that the density of the aqueous stream at the outlet is almost same as the density of the feed stream as the oxidant quantity is negligible with respect to the feed solution flow rate.

$$\rho_0 = \frac{F_i \rho_i + F_{r_i} \rho_{r_i}}{F_i + F_{r_i}} \quad F_o = F_i + F_{r_i} \qquad (2.24)$$

Cross-sectional area and volume of the reactor are computed as

$$A = \frac{\pi D_r^2}{4} \quad V = hA \qquad (2.25)$$

2.4.6.2 Computation of root mean square velocity gradient (G) in the coagulator/flocculator

The root mean square velocity gradient (in s^{-1}) in the coagulator–flocculator was computed as

$$G_1 = 1{,}000 \sqrt{\frac{\rho_{QM_o}}{V_{QM} \nu_{QM}}} \qquad (2.26)$$

2.4.6.3 Computation of average flock size (d_{QM}) in the coagulator–flocculator unit

Diameter of flock particles in the coagulator–flocculator is computed using the empirical relation

$$d_{QM_f} = \left(\frac{3}{2E_1 G_1 C_{QM_{floc}} t} \right)^{\frac{1}{3}} \qquad (2.27)$$

2.4.6.4 Computation of flow rate and concentration of coagulant

The flow rate of the coagulant is determined using a factor considering the stoichiometry of the reaction.

For low and stoichiometric feed rate of the coagulant dose:

$$F_{c_i} = f_2 F_{QM_i}, \text{ where } f_2 < 1$$

$$C_{c_i} = \frac{F_{QM_i} E_{As} C_{Ao} M_c}{M_{As} F_{ci} E_c} \qquad (2.28)$$

where $E_{As}:E_c = 1$ mg:50 mg; and M_{As}, M_c, and M_{floc} are molecular weights of arsenic, coagulant, and average molecular weight of the flocks, respectively.

Assuming negligible change in density of the aqueous stream as it passes from the inlet of the oxidizer unit to the outlet of the filter unit,

$$\rho_{QM_o} = \frac{F_{QM_i} \rho_{QM_i} + F_{c_i} \rho_{c_i}}{F_{QM_i} + F_{c_i}}, \qquad (2.29)$$

$$F_{QM_o} = F_{QM_i} + F_{c_i}, \quad F_{QM_i} = F_o, \quad \rho_{QM_i} = \rho_0 \qquad (2.30)$$

Area and volume of the coagulator:

$$A_{QM} = \frac{\pi D_{QM}^2}{4} \text{ and } V_{QM} = h_{QM} A_{QM} \qquad (2.31)$$

2.4.6.5 Determination of settling velocity and superficial velocity in sedimentation unit

When $d_P < 1$mm and $N_{Re} < 1$ where $d_P = d_{SM_f}$

$$U_1 = \frac{(\rho_S - \rho_L) g d_P}{18 \mu_L} \qquad (2.32)$$

$$N_{Re} = \frac{d_P^3 \rho_L (\rho_S - \rho_L) g}{18 \mu_L^2} \qquad (2.33)$$

When $d_P > 1$mm and $N_{Re} > 1$

$$U_2 = \frac{(S_P - 1)^{0.8} g^{0.8} d_P^{1.4}}{10 \, v_L^{0.6}} \qquad (2.34)$$

$$N_{Re} = \frac{d_P \rho_L U_2}{\mu_L} \qquad (2.35)$$

where $v_L = \frac{\mu_L}{\rho_L}$ and $S_P = \frac{\rho_S}{\rho_L}$

where U_1 and U_2 are the particle settling velocities. ρ_L and ρ_S are the densities of the particles and the aqueous solution. μ_L (cp) is the viscosity of the aqueous solution in the sedimentation unit.

$$Q_0 = F_0 \left(1 - \frac{C}{C_u} \right) \qquad (2.36)$$

where F_o is the input flow rate of the aqueous solution. Q_o is the overflow rate. C and C_u are the concentration of the flocks and sludge. A_d is the area of the sedimentation unit.

$$V_{actual} = \frac{Q_0}{A_d} \text{ or } V_{actual} = F_0 \frac{\left(1 - \frac{C}{C_u}\right)}{A_d} \quad (2.37)$$

where $A_d = \frac{\pi D_s^2}{4}$ and D_s is the diameter of the sedimentation unit. V_{actual} is the actual upward velocity of overflow water.

Check if $V_{actual} < U_i$, U_i means U_1 or U_2

If false, then increment the value of diameter, D_S of the sedimentation unit, and recalculate V_{actual}.

If true, proceed below to calculate efficiency:

$$\eta = 1 - \left[1 + n\left(\frac{U}{V_{actual}}\right)\right]^{-\left(\frac{1}{n}\right)}, n = 0 \text{ or } \frac{1}{8} \text{ or } \frac{1}{4} \text{ or } \frac{1}{2} \text{ or } 1 \quad (2.38)$$

$$V_{desired} < U$$

2.4.6.6 Determination of the filtration pressure drops due to filter cake and filter medium

$$\varepsilon = \varepsilon_0 \left(1 - 0.39t^2 - 0.45t\right) \quad (2.39)$$

$$L = 0.34t^{0.5} + 0.001 \quad (2.40)$$

$$V_F = V_0 F_0 (1 - 0.003Lt) \quad (2.41)$$

$$(-\Delta P_c) = 180 \mu L \frac{V_F (1 - \varepsilon)^2}{d_p^2 \varepsilon^3} \quad (2.42)$$

$$-\Delta P = (-\Delta P_c) + (-\Delta P_f) \quad (2.43)$$

where ε is the porosity of the filter cake, L is the cake thickness (m), and $-\Delta P_f$ is the pressure drop through the filter medium.

2.4.6.7 Effects of the operating parameters

A model profile of predicted arsenic concentration of treated water as depicted in Figure 2.6 shows around 91–92% removal of arsenic from the aqueous phase at steady state. Over an initial period of 30 min, experimental values are far below the model predicted values. This deviation of the initial phase

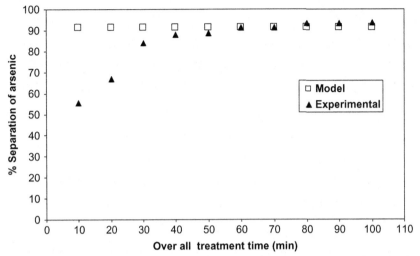

Figure 2.6 Arsenic concentration profile of the treated water. Experimental conditions: oxidant (KMnO$_4$) conc. 15 ppm; coagulant (FeCl$_3$) conc. 30 ppm; arsenic conc. of feed water 1.0 ppm; feed rate 1.32 dm^3/min; pH = 7.6; temperature = 305 K. Other conditions as in Table 2.1.

is attributed to the unsteady state of the whole plant during this phase. The deviation, however, is gradually smoothed out and the experimental findings are observed to corroborate well with the model predictions. The overall correlation coefficient is found to be 0.9895. This arsenic removal efficiency of around 91–92% is achieved for a feed concentration of 1 ppm (mg/L) of arsenic, coagulant (FeCl$_3$) concentration of 30 mg/L and oxidant (KMnO$_4$) dose of 15 mg/L. In batch studies, however, a higher percentage of removals (97–98%) have been reported by Shen [10]. The model assumes separation of arsenic basically through enmeshment and adsorption of arsenic onto the metal hydroxides but other mechanisms like formation of precipitates, coprecipitates, and mixed precipitates might also be active during the initial unsteady phase, resulting in a separation higher than model-predicted ones after the system attains steady state as shown in Figure 2.6.

Close agreement of the model predictions with the experimental findings only suggests that the model assumptions are largely correct and the model is capable of satisfactorily predicting the performance of the plant.

Model predictions may also be compared with experimental findings while studying the effects of major operating variables like oxidant dose, coagulant dose, and feed concentration. The optimum pH can always be arrived at experimentally.

2.4.6.8 Effect of pH

Figure 2.7 shows that removal of arsenic sharply rises from 60 to 90% as pH rises from 3.0 to 7.0. From pH 7.0 to pH 10.0 arsenic removal efficiency still increases (up to 92%) albeit marginally. However, beyond a pH level of 10.0, percentage removal of arsenic exhibits a negative correlation with pH. The optimum pH appears to be within a range of 7.6–8.0.

Ferric salts used as coagulants precipitate following the reaction:

$$Fe(OH)_3(s) \longrightarrow Fe^{3+} + 3OH^- \tag{2.44}$$

This equation shows that solubility of ferric hydroxide decreases with the increase of pH. Thus as pH increases, precipitation of ferric hydroxide and hence that of arsenic increases. Beyond a value of 10.0, the effect of higher pH becomes rather antagonistic on arsenic removal; this can be traced to the fact that at such pH levels, hydroxide begins to dissociate to form the soluble anion $Fe(OH)_4^-$ following the reaction

$$Fe(OH)_3(s) + OH^- = Fe(OH)_4^- \tag{2.45}$$

From the solubility product concept also, this precipitation can be correlated to pH. The solubility product constant (K_{sp}) for ferric hydroxide can be expressed as $K_{sp} = [Fe^{3+}][OH^-]^3$. Taking the logarithm of both sides the following relation is derived:

$$\log K_{sp} = \log\left[Fe^{3+}\right] + 3\log[OH^-]$$
$$\log\left[Fe^{3+}\right] = \log K_{sp} - 3\log K_w - 3pH \tag{2.46}$$

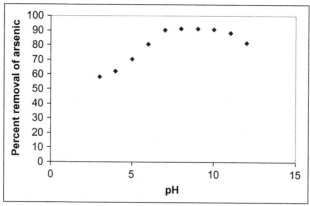

Figure 2.7 Effect of pH on % removal of arsenic. Experimental conditions: oxidant (KMnO₄) conc. 15 ppm; coagulant (FeCl₃) 30 ppm; arsenic conc. of feed water 1.0 ppm; feed rate 1.32 dm³/min; temperature = 306 K. Other conditions as in Table 2.1.

where K_w is the dissociation constant for water (at 298 K) $= [H^+][OH^-]$ $= 10^{-14}$.

Taking the solubility product constant of $Fe(OH)_3$ as 4×10^{-38} at 298 K, residual free ion concentration for iron may be expressed as

$$\log [Fe^{3+}] = 4.602 - 3pH \qquad (2.47)$$

Arsenic coprecipitates with $Fe(OH)_3$ as $As(V) - Fe(OH)_3$ flock, and this precipitation of arsenic is directly related to precipitation of $Fe(OH)_3$. In general ferric hydroxides precipitation causes an almost equal or even greater percentages of arsenic precipitation [13]. The optimum pH is found to be 7.6 for drinking water.

2.4.6.9 Effect of oxidant dose

Figure 2.8 shows a very pronounced effect of oxidant dose on arsenic removal from water. For a feed concentration of 1 ppm, percentage removal exhibits a strong positive correlation with oxidant dose for the investigated range of 2–25 ppm $KMnO_4$. Over the range 2–10 ppm, removal exhibits high sensitivity to the oxidant dose. Beyond this range (10 ppm) further increase of oxidation dose does not result in much improvement in

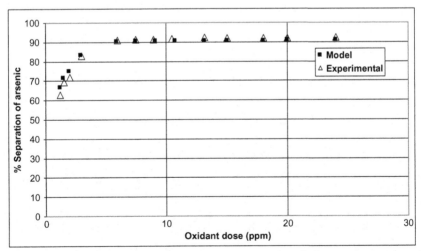

Figure 2.8 Effect of oxidant concentration on percent removal of arsenic. Experimental conditions: oxidant $KMnO_4$; coagulant ($FeCl_3$) conc. 30 ppm; arsenic conc. of feed water 1.0 ppm; feed rate (dm^3/min) 1.32; pH in the oxidation unit 6.5; pH in the coagulator 7.6; temp 305 K.

separation as is evident from the model as well as from experimental observations. From the 10 to 15 ppm level of oxidant dose, a marginal increase in
percentage removal is observed and the optimum oxidant dose is attained at
the 15 ppm level. Though data on the direct effect of oxidant dose on percentage removal of arsenic is virtually nonexistent, the effect of the oxidation
state of arsenic on its removal in chemical precipitation has been observed by
many researchers [6,9].

Improvement of separation efficiency up to 15 ppm oxidant dose is due
to the fact that As(V) separates with a much higher degree than As(III) in the
presence of the $FeCl_3$ coagulant [10]. Anionic species get involved in surface
chemical reaction with the binding sites of the adsorbent ferric hydroxides
surfaces, which remain predominantly positive up to a pH level of 7; As(III)
species being uncharged at this pH level cannot be bound to the adsorbent
surface. This explains better percent removals of As(V) than As(III). When
the medium reaches an oxidant concentration of 15 ppm, conversion of
As(III) into As(V) reaches the highest level and thus a further increase of
oxidant fails to raise the separation efficiency.

2.4.6.10 Effect of coagulant dose

Figure 2.9 shows the effect of the coagulant concentration ($FeCl_3$) on arsenic
removal. The figure indicates that up to the level of a 20 ppm coagulant
dose, separation of arsenic rises sharply with an increase of coagulant dose.
Experimental data follow the model closely. However, as the coagulant dose
reaches the 20 ppm level, the curve flattens, indicating less sensitivity of
removal efficiency to coagulant concentration beyond 20 ppm. However,
up to a 30 ppm concentration of coagulant, separation efficiency still rises
with the coagulant dose albeit marginally and settles at around a 91–92%
level for a 30 ppm coagulant dose.

The observed increase of percentage separation of arsenic with the
increase of coagulant dose is attributed to the increase of surface area
and active sites of the precipitating metal hydroxides onto which arsenic
gets adsorbed and coprecipitates. The benefits of iron hydroxide are
directly related to its concentration. It is found that [13] iron hydroxides
remove equal or greater percentages of soluble metal from aqueous solution. Moreover, iron coagulants ($FeCl_3$) dramatically reduce solubility of
arsenic, enhancing its precipitation. This rising trend of metal separation
efficiency with increase of coagulant dose continues until the hydroxide
surfaces onto which enmeshment and adsorption of metals takes place
become saturated.

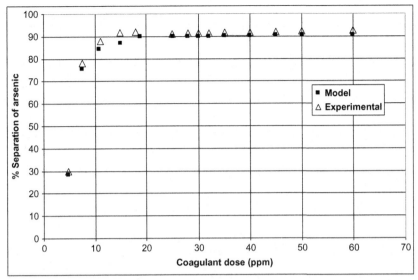

Figure 2.9 Effect of coagulant dose on % separation of arsenic. Experimental conditions: coagulant $FeCl_3$; oxidant $KMnO_4$ conc. 15 ppm; Arsenic concentration of feed water 1.0 ppm; feed rate 1.32 dm^3/min; pH (coagulator) 7.6; temp 305 K.

2.4.6.11 Effect of feed concentration

As depicted in Figure 2.10, over a feed concentration range of 0.2–2.0 ppm, removals are found to be independent of initial feed concentrations for excess coagulant (300 mg/L) and oxidant doses (80 mg/L). Such feed concentration behavior may seem to be contrary to the theoretical expectation that when a coagulant dose is not in excess, there is room for the surface site concentration to be controlling.

In the present model, adsorption–precipitation of arsenic is the dominant mechanism of arsenic removal from the aqueous phase. The extent of the adsorption is directly proportional to the concentration of the adsorbing surface sites, which in turn depends on the concentration of the coagulant. In the adsorption–precipitation mechanism, either concentration of the adsorbing surface sites or the concentration of the target solute (colloids) that act as nuclei in precipitate formation may be controlling. To decouple the effect of surface site concentration (i.e., coagulant dose) while studying the effect of initial feed concentration, the coagulant is used in excess. Thus for the low feed concentration range (0.5–2.0 mg/L) when the question of saturation of adsorption surface sites does not arise, percentage removal of arsenic is found to be independent of the initial feed concentration.

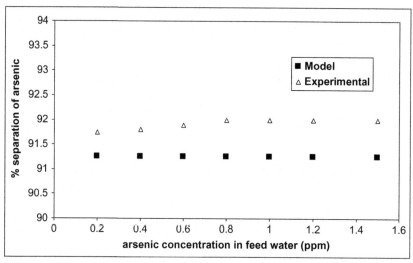

Figure 2.10 Effect of initial feed concentration on percentage separation of arsenic. Experimental conditions: oxidant (KMnO$_4$) conc. 80 ppm; coagulant (FeCl$_3$) conc. 300 ppm; feed rate 1.32 dm^3/min; pH 7.6; temp 304 K. Other conditions as in Table 2.1.

2.4.7 Performance of the system and the model

The dynamic mathematical model is developed to predict, a priori, performance of a physico-chemical arsenic separation plant. The model is validated against the experimental data using parameter values determined either experimentally or through standard empirical relations. Separation efficiency of 91–92% is achieved in the scheme. The parameters like coagulant dose, oxidant dose, and feed concentration are found to have a significant impact on arsenic removal efficiency. The results indicate high sensitivities of separation efficiency to the coagulant and oxidant doses over the lower concentration ranges. pH is also found to have a pronounced effect on separation efficiency of arsenic from water. Over the lower range (pH 4.0–7.0) percentage removal of arsenic is very strongly dependent on pH. As pH increases from 7.0 to 8.0, percentage removal of arsenic still increases, albeit marginally. However, the effect of pH turns antagonistic beyond a value of 10.0. The optimum pH value is thus found to be within the range of 7.6–8.0. Separation of arsenic in the process depends on a number of interdependent phenomena like formation of precipitates, coprecipitates, mixed precipitates, adsorption of inorganic arsenic species onto the metal hydroxides, enmeshment, and settling. Because it is very difficult to experimentally uncouple the effect of one phenomenon from the other, the initial wide gap between model predictions and experimental findings cannot be captured mathematically. The initial phase deviations of model predictions from experimental findings can only be traced to kinetic limitations, which can be solved through more rigorous kinetic study. Barring this initial phase, simulation results, however, agree well with the experimental findings (overall correlation coefficient being of the order of 0.9895). An automatic mechanism has also been incorporated into the model to automatically adjust the coagulant and oxidant doses based on optimum doses arrived at through several past studies in the event of fluctuation of feed concentration. The Visual Basic software developed is user-friendly, and the performance of such an arsenic separation plant can very quickly be analyzed under different operating conditions. Such a simulation developed for the first time is expected to be very useful in enhancing the confidence level of design and operation of a full scale physico-chemical arsenic separation plant.

2.5 OPTIMIZATION AND CONTROL OF TREATMENT PLANT OPERATIONS

For successful operation of a physico-chemical plant, optimization of the process variables and continuous monitoring are very essential. A Visual Basic software tool as that developed by Pal et al. [14] can be very effective in such optimization and control. Such user-friendly software in a well-known Microsoft Excel environment can be used for both optimizing operational conditions as well as for real-time monitoring during operation of the plant. Through visual graphics, this can permit very quick performance analysis of the individual units as well as the overall process. Flexibility in input data manipulation and capability of optimization of the major operating variables are the other expected advantages of such software. Such simulation software is very useful in raising the level of confidence in designing and operating arsenic separation plants. It appears that the most appropriate physico-chemical arsenic removal scheme should include in sequence one oxidation unit or reactor (with only $KMnO_4$ as oxidant), a coagulator or slow-mixing unit (with $FeCl_3$ as coagulant and provision for controlling pH), a flocculator or quick-mixing unit, and a sedimentation unit followed by one filtration unit like the sand filtration bed. Thus the software is based on a scheme that includes these units.

2.5.1 Development of the optimization and control software

The model equations for the integrated physical and chemical processes as developed in Sections 2.4.5.2–2.4.5.8 are used in developing the optimization and control software. The physico-chemical model parameters are determined either experimentally or by using standard mathematical relations available in the literature. Subsequently, an appropriate numerical solution technique is chosen and the algorithm is developed for the solution of the model equations. In the final step, the software is validated through experimental investigation and comparison between the model-predicted values and the experimental findings.

2.5.2 The overall procedure of computation and output generation

The overall procedure of computation and graphical output generation consists of the following steps.

1. A database containing initial parameter values is defined.
2. A solution of temporal derivatives is done by calling a Runge–Kutta–Fehlberg subroutine using an initial value database.
3. Physico-chemical model parameters are computed using standard theoretical correlations or through regression. Among the parameters, the time-dependent parameters are continuously updated in their respective databases until convergence. The other time-independent parameters are stored as constants in their respective databases.
4. The initial database is updated through step 2.
5. Comparison is done in the next step for set error tolerance and steps 1 through 4 are repeated until convergence.
6. The final values of the dependent variables thus obtained are then stored separately in their respective databases of different units.
7. Desired preset graphical outputs are then generated using the databases.

2.6 THE NUMERICAL SOLUTION SCHEME AND ERROR MONITORING

Other than simple algebraic equations, the model involved a number of coupled ordinary differential equations. For numerical solution of the coupled differential equations, the Runge–Kutta–Fehlberg method is used. The integration procedure incorporated an automatic integration step size adjustment mechanism. The maximum permissible relative error is set at 0.01 and all computations are carried out within this tolerance limit.

2.6.1 Software description

The ARSEPPA simulation software has been written in Visual Basic [14]. This is an add-in in Microsoft Excel. The user-friendly menu-driven program is capable of producing the output through visual graphics. The overall process consists of five different units, namely reactor or oxidizer, coagulator or quick-mixing unit, flocculator or slow-mixing unit, a sedimentation unit, and a filter unit. We can analyze the performance of the individual units as well as the overall process applying the software. The salient features of data input, data output, method setting, input data updating, and screen placement are illustrated in Figures 2.11 through 2.22.

The general data sheet as shown in Figure 2.11 appears when the software is run. It incorporates a user guide under the **Show Tips** option. The **Screen Placement** option permits visualization of different windows in different styles like **Tile Cascade**, **Horizontally**, **Tile vertically**, and so on. The **Choose Simulation Mode** option permits performance analysis of either individual units or the overall process as a whole. The **Data Handling Method** option incorporates the provision for setting the parameters of the input data sheet. Unless a new method is set up, the simulation runs by default using the set parameters. The **View** tab permits checking the saved data sheets under a specified data sheet number. **Clear All** helps to rewrite a new data sheet and erasing the existing one. **Update** saves the newly created data sheet.

Figure 2.12 shows the main window tool bar. The tool bar contains the icons of all the units', viz., reactor, quick mixing unit, slow mixing unit,

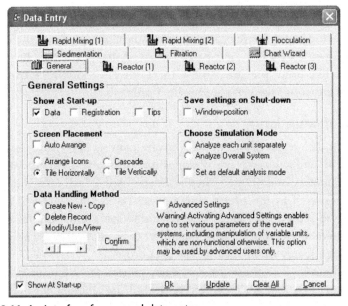

Figure 2.11 An interface for general data entry.

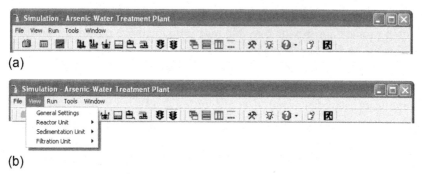

Figure 2.12 The user interface view of the tool bar.

sedimentation unit, and filtration unit. The tool bar provides for two separate tabs, **Start Simulation** and **Stop Simulation**. To run the simulation, select the desired unit and then click **Start Simulation**. Simulation results are displayed graphically. To get the results sheet, select the **Grid Data** menu. Graphical simulation results are obtained both in multiwindow fashion as well as in cascade style. Using the appropriate tools of the chart sheet tool bar, output can be printed or saved. The tool bar incorporates facility of graph editing. The grid data values can be directly transferred to an Excel sheet for generating secondary graph sheets. To analyze the overall system for performance, select the **Run** tab first, then the **Overall System** and **Start Simulation** options sequentially. The tool bar also has the provision for file handling under the name **Disk Utility Station**. You can create or remove a folder and delete or move a file using this tool. From the **File** tab, you can open a new run sheet or an old saved sheet. The tab also includes functions like print, preview, and so on. Different tools like **Export data sheet** to Excel or vice versa are in the **Tools** tab.

2.6.2 Software input

The input data required to run the software consists of physical dimensions of each unit and its auxiliary provisions (like stirrer, etc.), kinetic data, operating parameters, and physico–chemical data. Under each unit, the relevant data are entered in the preset item boxes. To save the entered data the **Update** tab is used. The **Data Entry** window has a provision for entering the data in different units. Editing the units can be easily done by pressing the **U** tab that appears when clicking the data boxes.

Figures 2.13–2.15 show the data entry pattern for the reactor.

Figures 2.16 and 2.17 exhibit how data are entered for the quick-mixing unit.

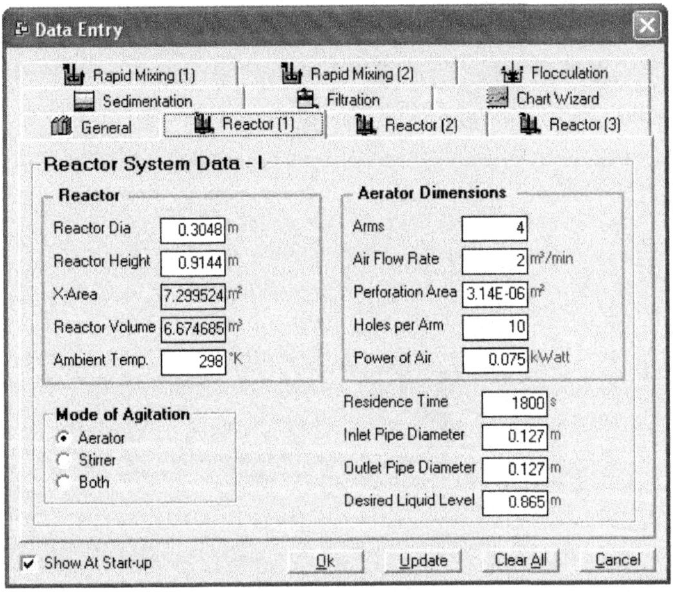

Figure 2.13 The first input data sheet of the reactor.

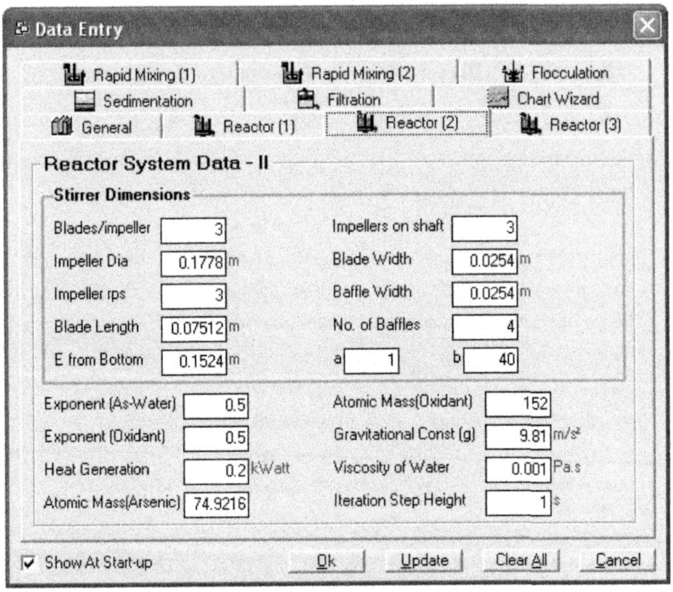

Figure 2.14 The second input data sheet of the reactor.

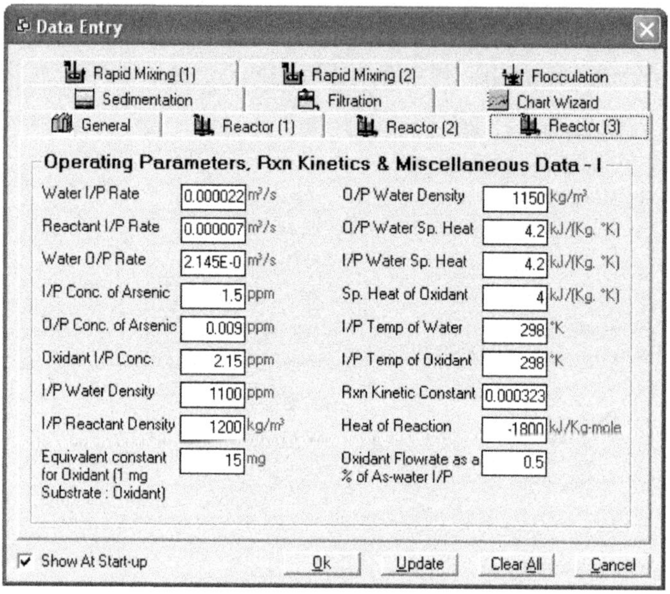

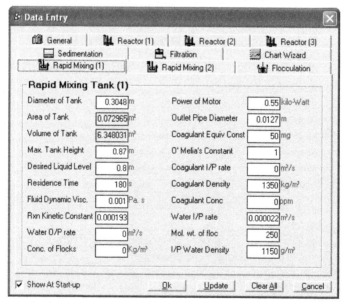

Figure 2.15 The third input data sheet of the reactor.

Figure 2.16 The first input data sheet of the quick-mixing tank (coagulator).

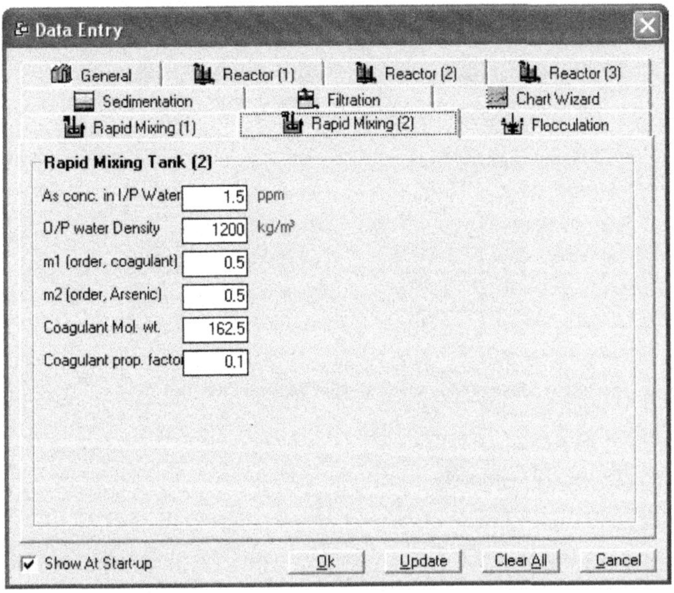

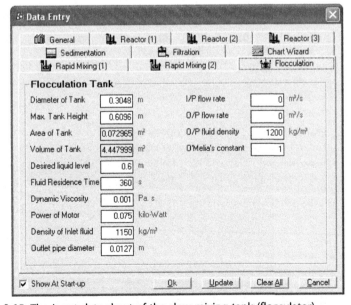

Figure 2.17 The second input data sheet of the quick-mixing tank (coagulator).

Figure 2.18 The input data sheet of the slow-mixing tank (flocculator).

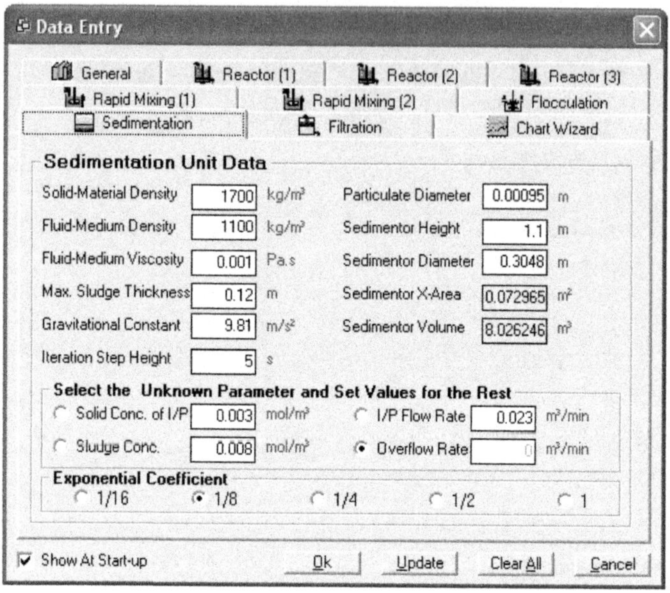

Figure 2.19 The input data sheet of the sedimentation unit.

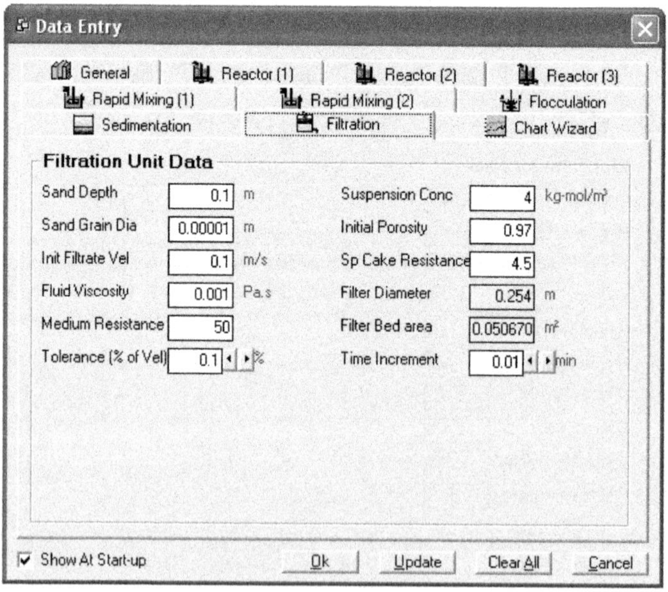

Figure 2.20 The input data sheet of the filter unit.

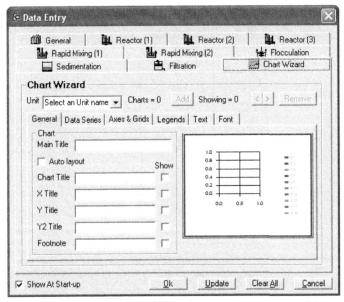

Figure 2.21 The chart wizard is used to see the performance of the different units.

2.6.3 Software output

Some typical output forms are shown in Figures 2.22 and 2.23.

Figure 2.23 exhibits how the software-predicted overall performance in terms of percentage separation of arsenic varies with the experimental findings as a major operating variable coagulant dose changes. A similar performance characteristic curve in Figure 2.22 shows the effect of coagulant dose.

2.6.4 Running the software

A typical set of conditions as presented in Table 2.3 can be used in running the software.

2.6.4.1 Software analysis

Despite extensive research work on several techniques of arsenic separation over the decades, millions of people in the developing countries, particularly in South East Asia, still continue to drink water highly contaminated with arsenic. There is still very limited confidence in design and operation of a physico-chemical treatment plant for arsenic removal from water as is evident from the operation of a very limited number of such plants. There is still doubt as to the effectiveness and economy of such a treatment plant. Though research abounds regarding physico-chemical separation of arsenic from

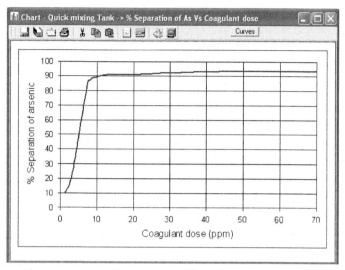

Figure 2.22 The output graph sheet generated from the program showing the effect of the coagulant dose.

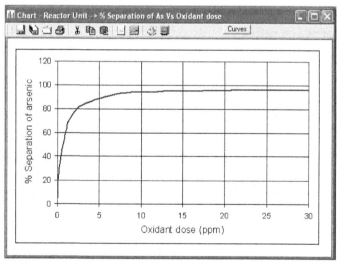

Figure 2.23 The output graph sheet generated from the program showing the effect of the oxidant dose.

drinking water, neither any systematic modeling and simulation work considered the most appropriate treatment scheme as outlined above, nor any software development work on the integrated processes concerned has been adopted yet in this vital area of drinking water purification.

Table 2.3 Typical Experimental Conditions and Model Parameters

Temp maintained in the units	298–305 K
Root mean square velocity gradient in the coagulator (G_1)	$800\ s^{-1}$
Root men square velocity gradient in the flocculator (G_2)	$70\ s^{-1}$
Feed water flow rate	$0.022 \times 10^{-3}\ m^3/s$
Arsenic concentration of the feed water	$1-2 \times 10^{-3}\ kg/m^3$
Oxidation rate constant	$3.23 \times 10^{-3} s^{-1}$
pH in the oxidation unit	5.5
pH in coagulator and flocculator	7–8
Overall settling rate constant	$1.93 \times 10^{-3} s^{-1}$
Coagulant concentration	$30 \times 10^{-3}\ kg/m^3$
Oxidant concentration	$15 \times 10^{-3}\ kg/m^3$

However this could be of great help in the full-scale design and operation of arsenic separation plants. The Visual Basic simulation software (ARSEPPA) based on dynamic mathematical modeling of all the systematically integrated physico-chemical processes of arsenic separation from drinking water is a result of integration of knowledge from computer software engineering and environmental engineering with chemical engineering. This permits a very quick performance analysis of the process units involved in the separation of arsenic from water. The major advantage of the user-friendly and menu-driven software is that it deals with a continuous process where the effects of all the major operating parameters on the effectiveness of arsenic separation can be observed. This in turn helps set the operating parameters at their optimum levels. This is a menu-driven add-in in the Microsoft Excel environment. It, therefore, does not require familiarity with any new environment. The software permits preanalysis manipulation of input data and visualization of the output in a familiar environment. Though developed for arsenic separation from drinking water, the software can be extended to separation of many other heavy metals like calcium, magnesium, iron, and lead from water.

2.7 TECHNO-ECONOMIC FEASIBILITY ANALYSIS

An economic evaluation was carried out for a plant with a capacity of 100,000 L/day (36,500 m^3/year) drinking water production (considering a community population of 5,000 and water consumption 20 L/capita). Since the plant is running 24 hr per day, the treated water produced per hour by the plant is 4,167 L. For scale up, cost assessment is done using a sixth-tenth power law, defined as:

$$\text{Scale-up cost} = \text{Lab scale cost} \times \left(\frac{\text{Scale-up data}}{\text{Lab scale data}}\right)^{0.6}$$

In capital cost ($) there are several cost items, such as civil or infrastructure cost, mechanical engineering cost, electro-technical and electricity cost, chemical cost, labor cost, maintenance cost, and overhead cost, which are considered per-year operational costs.

In the present study, Building construction, different types of tank construction (main feed tank, oxidation tank, coagulation tank, flocculation tank, sedimentation tank, sand filtration tank, and stabilization tank), and drainage system are involved in civil or infrastructure cost. Mechanical engineering cost is based on different types of pipe cost, stirrer cost, valve cost, and fittings cost.

High pressure pump cost, feed pump cost, agitator motor cost, rotameter cost, pH probe cost, and other electrical lines fitting cost are termed the electro-technical cost.

Electricity cost is one of the operational cost parameters and the cost has been calculated on the basis of electricity consumption per year by the different types of pump, agitator motor, any digital meter, and electric line fittings.

Different types of chemical cost also are involved in operational cost. $KMnO_4$ is used as an oxidant agent where $FeCl_3$ is used as a coagulant. The optimum dose of $KMnO_4$ and $FeCl_3$ are 15 and 30 mg/L, respectively. Two types of coagulants (ferric sulfate and calcium hydroxide) are used in the stabilization process, with an optimum dose of 250 mg/L of ferric sulfate and 500 mg/L of calcium hydroxide. Labor cost, maintenance cost, and overhead cost are also involved in operational cost. The calculated cost (capital and operational cost) is shown in Table 2.4.

The cost assessment is based on the annualized investment cost and annualized operational cost. Annualized capital cost was computed by the following relationship:

$$\text{Annualized capital cost} = \left(\frac{\text{Total capital}(\$) \times \text{Cost recovery factor}}{\text{Water flux per year}(m^3/\text{year})}\right)$$

The cost recovery factor was dependent on plant project life ($n = 20$ years) and interest ($i = 8\%$) and it can be calculated by the following equation:

$$\text{Cost recovery factor} = \left(\frac{i(1+i)^n}{(1+i)^{n-1}-1}\right)$$

Table 2.4 Capital and Operating Cost of a 100,000 L/Day Capacity Water Treatment Plant

Cost parameters Capital cost *Cost for civil infrastructure*	No of equipment with specification	Cost value cost ($)
Building construction	100 m² space	3,300
Concrete tank construction	5 (concrete tank, 5,000 L capacity)	4,200
Drain construction	100 m long	400
Electro-technical cost		
Aeration pump	1	100
Feed pump	1 (Submersible pump)	416
Low pressure pump cost	2 (centrifugal pump)	200
Agitator motor cost	3	3,000
pH probe	2	100
Mechanical engineering cost		
Cost for main feed pipe	80 m long and 0.1 m dia.	3,000
Stirrer cost	3	300
Valves and pipe fittings		300
Total cost		15,316

Operating cost		Cost ($/Year)
Electricity cost	Power consumption– 36,000 Kwh/year	3,000
Chemical cost for filtration		
$KMnO_4$ cost	547.5 kg/year (bulk cost: 9.2 $/kg)	5,037
$FeCl_3$ cost	1095 kg/year (bulk cost: 8.4 $/kg)	9,198
Chemical cost for stabilization		
Ferric sulfate	50 kg/year (bulk cost: 10 $/kg)	500
Calcium hydroxide	100 kg/year (bulk cost: 6.7 $/kg)	670
Sand cost	1 ton	100
Labor cost	No. of operators 3 (1000 $/month)	12,000
Maintenance charge	2% of capital investment	306
Overhead charge	2% of capital investment	306
Total cost		**31,117**

Again, annualized operational cost can be computed by the following equation:

$$\text{Annualized operational cost} = \left(\frac{\text{Total operational cost}(\$/\text{year})}{\text{Water flux per year}(m^3/\text{year})} \right)$$

Thus annualized cost for production of 1000 L of drinking water is the summation of annualized capital cost and annualized operating cost $= 0.046 + 0.85 = 0.89$ \$.

NOMENCLATURE

F_i, F_0	volumetric flow rates (m^3/s) of the feed and treated water, respectively
F_r	volumetric feed rate (stoichiometric) of oxidant
ρ_i, ρ_0	densities (kg/m^3) of water
ρ_r	density of oxidant
A	reactor cross-sectional area
h	liquid level in the reactor
C_A, C_A	concentration $(kmol/m^3)$ of As(III) at the inlet and As(V) at the outlet of the reactor
C_r	oxidant concentration $(kmol/m^3)$
k	second order reaction (oxidation) rate constant $(mol^{-1} - s^{-1})$
n_1, n_2	kinetic constants
V	volume of the reactor (m^3)
C_{r_i}, C_r	initial and instantaneous concentration of the oxidant $(kmol/m^3)$
ρ_{QM_i}, ρ_{QM_o}	densities (kg/m^3) of the inlet and outlet aqueous solutions in the coagulator-flocculator
ρ_{C_i}	density of the coagulant (kg/m^3)
A_{QM}	area of the coagulator/flocculator (m^2)
h_{QM}	liquid level in the coagulator (m)
F_{QM_i}, F_{QM_o}, F_{C_i}	flow rates (m^3/s) of the feed water, treated water, and coagulant
$C_{QM_{A_i}}$, C_{QM_A}	concentrations $(kmol/m^3)$ of arsenic at the inlet and outlet
C_c	coagulant concentration $(kmol/m^3)$ in the coagulator
V_{QM}	volume of the coagulator (m^3)
K_{QM}	assumed overall second order rate constant $(mol^{-1} - s^{-1})$ of arsenic flocculation, adsorption, enmeshment, and settling
m_1, m_2	reaction kinetic constants
$C_{QM_{floc}}$	concentration of the flock $(kmol/m^3)$
M_{floc}	average molecular weight of flock $(kg/kmol)$
$G_1(s^{-1})$	average root mean square velocity gradient in the coagulator-flocculator
$D_{QM\,f}$	average diameter of the flock particles in the coagulator-flocculator (m)
F_o	volumetric feed rate of aqueous solution in the sedimentation unit (m^3/s).
C	flock concentration of the solution $(kmol/m^3)$
A_d	sedimentation unit area (m^2)
C_u	sludge concentration or flock concentration $(kmol/m^3)$
U	average settling velocity of the flock particles (m/s)

$\dfrac{dz}{dt}$	sedimentation rate (m/s)
V_F	volume of filtrate (m^3)
A	area of the filter bed
α	specific cake resistance (m/kg)
W	solid concentration of the water to be filtered
μ	viscosity of the aqueous solution at the inlet of the filter unit
R_m	filter medium resistance
$(-\Delta P)$	pressure drop through the filter medium and filter cake (N/m^2)
M_{As}, M_r, M_c, M_{floc}	molecular weights of arsenic, oxidant, coagulant, and average molecular weight of the floc(kg/kmol)
D_r	reactor diameter (m)
V	volume of the reactor (m^3)
h	height of the reactor (m)
P	power in Nm/ s
μ	dynamic viscosity of water in Pa-s
V	volume of coagulator/ flocculator in m^3
D_{QM_f}	diameter of floc particles in the coagulator-flocculator
D_{QM}	coagulator diameter (m)
V_{QM}	volume of the coagulator (m^3)
h_{QM}	height of the coagulator (m)
D_P or D_{SM_f}	diameter of the floc (m) in the flocculator
U_1, U_2	particle settling velocities
ρ_S, ρ_L	densities of the particles and the aqueous solution
μ_L	viscosity (cp) of the aqueous solution in the sedimentation unit
F_o	input flow rate of the aqueous solution
Q_o	overflow rate
C, C_u	concentration of the floc and sludge
A_d	area of the sedimentation unit
V_{actual}	actual upward velocity of overflow water
N_{Re}	Reynolds number
g	gravitational constant (m/s^2)
S_p	specific density
ν_L	kinematic viscosity (m^2/s)
ε	porosity of the filter bed
L	cake thickness(m)
ΔP_f, ΔP_c, ΔP	pressure drop across filter medium, filter cake, and total pressure drop across the bed

REFERENCES

[1] Robins RG. The solubility of metal arsenates. Proc. MMIJ-AIME Joint Meeting, Tokyo, Paper D-1-3; 1981. p. 25–43.

[2] Khoe GH, Emmett MT, Zaw M, Prasad P. Removal of arsenic using advanced oxidation processes. In: Young CA, editors. Minor metals; 2000.

[3] Kartinen EO, Martin CJ. An overview of arsenic removal processes. Desalination 1995;103:79–88.

[4] Wang Q. Theoretical analysis of Brownian heterocoagulation of fine particles at secondary minimum. J Colloid Interface Sci 1991;145:305–13.

[5] Hiemenz PC, Rajagopalan R. Principles of colloid and surface chemistry. 3rd ed. New York: Marcel Dekker; 1997.

[6] Israelachvili JN. Intermolecular and surface forces. 2nd ed. New York: Academic Press, Inc; 1992.

[7] Hering JG, Elimelech M. Arsenic removal by ferric chloride. J Am Water Works Assoc 1996;88(4):155–67.

[8] Cheng RC, Liang S, Wang HC, Beuhler MD. Enhanced coagulation for arsenic removal. J Am Water Works Assoc 1994;86(9):79–90.

[9] Pal P, Ahamad Z, Pattanayak A, Bhattacharya P. Removal of arsenic from drinking water by chemical precipitation—a modeling and simulation study of the Physical-chemical processes. Water Environ Res 2007;79(4):357–66.

[10] Shen YS. Study of arsenic removal from drinking water. J Am Water Works Assoc 1973;65(8):543–7.

[11] Stumm W, Morgan JJ. Aquatic chemistry. New York: Wiley; 1980.

[12] O'Melia CR. Coagulation and flocculation. In: Weber WJ, editor. Physicochemical processes for water quality control. New York: Wiley-Interscience; 1972.

[13] Edwards M, Benjamin MM. Regeneration and reuse of iron hydroxide adsorbents in treatment of metal-bearing wastes. J Water Pollut Control Fed 1989;61(4):481–90.

[14] Pal P, Ahamad Z, Bhattacharya P. ARSEPPA: visual Basic software tool for arsenic separation plant performance analysis. Chem Eng 2007;129:113–22.

CHAPTER 3

Adsorption Method of Arsenic Separation from Water

Contents

Groundwater Arsenic Remediation
http://dx.doi.org/10.1016/B978-0-12-801281-9.00003-5

71

3.1 INTRODUCTION

Adsorption is a phenomenon of concentration of certain solutes from a solution or mixture onto the surface of a solid by virtue of the unbalanced forces of attraction of the atoms at the surface of the solid. Such attractive forces of the atoms at the interior of the solid get balanced through interactions with forces of other atoms in the lattice. Adsorption may be physical or chemical. Physical adsorption is due to van der Waals forces of attraction only, whereas chemical adsorption involves chemical interactions of the solute molecules with the solid surface of the adsorbent. Adsorption is a very general phenomenon and demands certain specific characteristics of the adsorbent for economical use in practical or industrial fields. These are selectivity, capacity, and life of the adsorbent material. In general when comparison is made with distillation (the standard separation process), an adsorption-based process is likely to be more economical than distillation if relative volatility of the two components of the mixture or solution falls below 1.2. Suitability of an adsorbent can often be assessed through measurement of Henry's constants or through direct chromatographic retention time measurements.

Adsorption isotherms are used for understanding the mechanism and quantifying the distribution of the adsorbate between the two phases (gas–solid or liquid–solid) of the adsorbate carrying fluid and the solid adsorbent at equilibrium during the adsorption process. Langmuir [1] had developed adsorption isotherms based on the concept of monolayer adsorption. Later, Brunauer et al. [2] developed adsorption isotherms known as BET isotherms based on the concept of multilayer adsorption, which could explain physical adsorption in a much better way. While Langmuir isotherm applies normally to low-pressure adsorption, BET isotherms describe adsorption under high-pressure conditions. Under high-pressure, low-temperature conditions, a larger number of the adsorbate molecules remain in contact with the surface of the adsorbents because of their low thermal energy, resulting in multilayer adsorption. Five basic types of adsorption isotherms as identified by BET are shown in Figure 3.1. These isotherms show

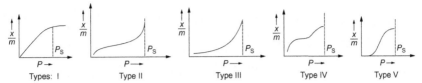

Figure 3.1 Various adsorption isotherms as classified by Brunauer et al. [2].

mass (x) of adsorbate per mass (m) of adsorbent against pressure (P), where P_s represents maximum saturation pressure.

Variation in adsorption behavior as exhibited in the five basic types of adsorption isotherms is mainly based on relative sizes of the sorbate molecules and the micropores of the adsorbent. Type I isotherm applies to situations when size difference between the micropore of the adsorbent and the sorbate molecules is not very significant. Thus, there is a definite saturation limit corresponding to complete filling of the micropores of the sorbent by the sorbate molecules. Adsorbents having wide ranges of pores within their structures exhibit type II and type III isotherms. In these adsorbents, continuous increase in adsorption capacity is observed with increase in pressure as capillary condensation takes place. Type IV isotherms represent adsorption on two distinct surfaces—one on the plane surface and the other on the walls of the pores where the pore diameter is much larger than the sorbate molecules.

3.2 ADSORPTION KINETICS

Quite a few empirical models have been developed to capture adsorption kinetics. Among these the most commonly used in sorption studies are the Langmuir [1], BET (Brunauer–Emmett–Teller) [2], Freundlich [3], and Redlich–Peterson [4] models. The Langmuir model is valid for monolayer adsorption on a surface with a finite number of adsorption sites of equal energy and is expressed as

$$A_e = \frac{A_{m,e} k_L C_e}{1 + k_L C_e} \tag{3.1}$$

where A_e and $A_{m,e}$, respectively, represent the amount of adsorption and the maximum amount of adsorption per unit of adsorbent used at equilibrium arsenic concentration, C_e. k_L is the equilibrium coefficient of the Langmuir model. The linearized form of this model is expressed as

$$\frac{C_e}{A_e} = \frac{1}{k_L A_{m,e}} + \frac{C_e}{A_{m,e}} \tag{3.2}$$

The Freundlich model is an empirical equation based on the assumption that the adsorbent has a heterogeneous surface composed of different classes of adsorption sites, and each site can be modeled by the following equation:

$$A_e = k_F C_e^{1/n} \tag{3.3}$$

where k_F is the equilibrium coefficient of the Freundlich isotherm and n represents the intensity constants of the adsorbents. The linearized form of this model is

$$\ln A_e = \ln k_F + \frac{1}{n} \ln C_e \tag{3.4}$$

The Redlich–Peterson model combines the Freundlich and Langmuir models, as well as the heterogeneity of the sorbent surface and a certain number of adsorption sites with the same adsorption potential. The equation by Redlich–Peterson is the following:

$$A_e = \frac{k_R C_e}{1 + a C_e^b} \tag{3.5}$$

where k_R is the equilibrium coefficient of the Redlich–Peterson isotherm and a and b represent equation constants. The linearized form of this model is

$$\ln \left(\frac{k_R C_e}{A_e} - 1 \right) = \ln a + b \ln C_e \tag{3.6}$$

The BET model provides coverage of the surface with adsorbate multilayers, and as active sites have different energies, multilayers can be formed in different parts of the surface and the BET equation can be expressed as

$$A_e = \frac{A_m k_{BET} C_e C_{Fl}}{(C_{Fl} - C_e) + (C_{Fl} + k_{BET} C_e - C_e)} \tag{3.7}$$

where k_{BET} is the equilibrium coefficient of the BET isotherm and C_{Fl} is the filled layer concentration. The linearized model of the BET isotherm is

$$\frac{C_e}{A_e(C_{Fl} - C_e)} = \frac{1}{A_m k_{BET}} + \frac{k_{BET} C_e - C_e}{A_m k_{BET} C_{Fl}} \tag{3.8}$$

The equilibrium constants (k_L, k_F, k_R, k_{BET}) and different constants (a, b, n) of the equations are calculated from the slope and the intersection of the line graph by plotting $\frac{C_e}{A_e}$ vs. C_e (Eq. 3.2), ln A_e vs. ln C_e (Eq. 3.4), $\ln\left(\frac{k_R C_e}{A_e} - 1\right)$ vs. ln C_e (Eq. 3.6) and $\frac{C_e}{A_e(C_e - C_{Fl})}$ vs. $\frac{C_e}{C_{Fl}}$ (Eq. 3.18).

3.3 ADSORBENTS USED IN ARSENIC REMOVAL

3.3.1 Synthetic activated carbon-based adsorbents

Activated carbon-based adsorbents have been developed using natural materials like coconut shell, almond shell, fertilizer slurry, palm tree cobs, petroleum coke, pine saw dust, bituminous coal, lignite, and fly ash.

3.3.2 Metal-based adsorbents

A very large number of metal-based adsorbents have been developed over the years through scientific research across the world. These metal-based adsorbents as reported in the literature may be listed as iron oxide coated sand, red mud, activated alumina, MnO_2, Z_r resin, iron(III)-loaded chelating resin, TiO_2, $FePO_4$ (amorphous), MnO_2-loaded resin, iron(III) oxide-loaded melted slag, oxisol, gibbsite, goethite, kaolinite, zirconium-loaded activated carbon, granular ferric hydroxide, ferrihydrite, mixed rare earth oxide, Portland cement, iron oxide–coated cement, hematite, feldspar, aluminum-loaded coral limestone, Fe(III) alginate gel, ferric chloride impregnated silica gel, titanium dioxide-loaded Amberlite XAD-7 resin, and iron(III)-loaded chelating resin.

3.3.3 Performance of the adsorbents

3.3.3.1 Activated carbon

Activated carbon produced from coconut shell, peat, and coal are found [5] to remove arsenic by 2.4, 4.91, and 4.09 mg/g carbon, respectively. With pretreatment by Cu(II) solution, removal capacity improves further. As(V) is more effectively removed from solution by using activated carbon with a high ash content. Adsorption of arsenic is found to decrease with increasing pH. The optimum pH is 6.0 for arsenic adsorption by pretreated carbon.

Using Ni-loaded activated carbon [6] around 95% arsenic can be removed from aqueous solution where insoluble Ni(II) arsenates formed on the surface of alumina facilitates arsenic adsorption. A coal-based mesoporous activated carbon was prepared by Wei–Guang Li et al. [7] for arsenic removal from low-temperature micropolluted water. At an adsorbent dose of 200 mg/L arsenic could be removed by more than 90%.

3.3.3.2 Activated alumina

Alumina has been in use as an adsorbent since the dawn of the twentieth century. First reported use of alumina in chromatographic purification of biological compounds dates back to 1901 [8].

Its commercial introduction was in 1932 by Alcoa Company [9] for adsorption of water. Since then it has traditionally been used as a desiccant. Subsequently its use has been diversified as a catalyst in chemical production systems and as an adsorbent of impurities in water treatment. By virtue of its abundance, high mechanical strength, large specific surface area, and ease of regeneration, activated alumina has been used not only in arsenic removal but also in fluoride removal and purification of water in general in point-of-use facilities. Sen and Pal [10] developed low-cost activated alumina with a large specific surface area of 335–340 m^2/g from gibbsite precursor through partial thermal dehydration and used in the successful removal of arsenic from contaminated water up to a level well below the World Health Organization (WHO)-prescribed maximum concentration limit of 10 μg/L.

3.3.3.3 Zeolites

Natural zeolites can also remove arsenic from aqueous solution but efficiency of activated carbon for removal of As is found [11] to be much higher than for natural zeolite. Activated carbon has been demonstrated to remove 60% As(V) and As(III), whereas natural zeolite has been found [12] to remove 50% As(V) and only 30% As(III). Compared to natural zeolites, synthetic zeolites are found [13] to remove arsenic to a much greater extent. Different porous nature as well as the Si/Al ratio of the adsorbent has great impact on the arsenic removal process. Over 70% arsenic is found [13] to be removed over a treatment time of 80 min.

3.3.3.4 Red mud

Red mud, a waste from the aluminum industry, has been found to remove almost 100% arsenic (initial concentration of 133 μmol/L) from water by applying 100 g/L red mud dose into the contaminated water [14,15]. Novel iron-based red mud sludge was used by Li et al. [16] as a cost-effective adsorbent for arsenic removal from aqueous solution. With an initial arsenic concentration of 0.5 mg/L, more than 90% arsenic could be removed at an adsorbent dose of 0.4 g/L.

3.3.3.5 Rice husk

Untreated rice husk has been observed [17] to remove both As(III) and As(V) by almost 100% from aqueous arsenic solution at a pH of 6.5, initial arsenic concentration of 100 μg/L, and adsorbent dose of 6 g/L.

3.3.3.6 Fly ash

Fly ash obtained from coal power stations can be used [18] to reduce arsenic concentration of aqueous solution from 500 to 5 ppb.

3.3.3.7 Hematite and feldspar

Hematite and feldspar have been used [19] effectively in the removal of arsenic in pentavalent form from aqueous solutions at different pH, temperatures, and adsorbent particle sizes. Uptake followed first-order kinetics and fitted the Langmuir isotherm. The maximum removal was 100% with hematite (pH 4.2) and 97% with feldspar (pH 6.2) at an arsenic concentration of 13.35 μmol/L.

3.3.3.8 Coated sand

A variety of treated and coated sand have been used in arsenic remediation [20,21], exploiting their highly porous structures. Manganese green sand, iron oxide-coated sand (IOCS-1, IOCS-2), and ion-exchange (Fe^{3+} form) resin columns have been used in arsenic removal from tap water. Batch studies of IOCS-2 demonstrated an organic arsenic adsorption capacity of 8 μg/g IOCS-2. High bed volumes (585 BV) and high arsenic removal capacity (5.7 μg/cm^3) are achieved by this resin. Nguyen et al. [22] also synthesized iron-oxide-coated sponge (IOCSp) for As(III) and As(V) removal and found that 1 g of IOCSp adsorbed about 160 μg of arsenic within 9 hr.

3.3.3.9 Nanoparticles

Cu[II] oxide nanoparticles have been used [23] as adsorbent for arsenic removal from water. High surface area of the nano size adsorbents adsorbs high amount of toxic arsenic from water. One hundred percent removal efficiency was achieved for initial concentration up to 200 μg/L within 3 hr of treatment at adsorbent dose of 2 g/L. Olyaie et al. [24] developed a cost-effective technique to remove arsenic contamination from aqueous solutions by calcium peroxide nanoparticles. Oxidation occurred within minutes and CaO_2 nanoparticles successfully removed arsenic at a natural pH range. Reported removal efficiency achieved is over 80% at an adsorbent dose of 40 mg/L in a 40-min contact time.

3.3.3.10 Laterite soil

Low-cost laterite soil has also been used by some researchers [25] as a natural adsorbent for arsenic removal. Under optimum adsorbent dose of 20 g/L and the contact time of 30 min, the laterite soil could remove up to 98% of total arsenic.

3.3.3.11 Portland cement

Portland cement as a low-cost adsorbent may be used [26] for the removal of arsenic from water. The adsorbent can remove up to 95% arsenate at an initial arsenate concentration of 0.2 ppm.

3.3.3.12 Iron-based adsorbent

Granular ferric hydroxide is found to be an effective adsorbent [27] for arsenic separation. At equilibrium, about 95–99% of the arsenic [V] is adsorbed when an arsenic [V] to granular ferric hydroxide ratio is maintained at 400 µg As/g GFH. A novel adsorbent (high iron-containing fly ash) for arsenic (V) removal from wastewater was developed by Li et al. [28]. The highest surface area of the adsorbent provides a high removal percentage of arsenic from wastewater. Around 96% of arsenic was removed at a dose of 60 g/L of adsorbent.

IOCSp has been used as an adsorbent and it is found to remove around 95% arsenic from a synthetic solution containing arsenic as high as 1000 µg/L [29]. Magnetic iron oxide nanoparticles from tea waste has been successfully synthesized [30] in removing almost 99% of arsenic by applying an adsorbent dose of 0.5 g/L. Synthetic magnetic wheat straw with different Fe_3O_4 content has also been found effective in arsenic removal [31]. This material can be used for arsenic adsorption from water, and can easily be separated by an applied magnetic field. Around 80% arsenic is found to be removed by applying 3 g/L adsorption dose in water.

3.3.3.13 Other adsorbents

Chio et al. [32] used low-cost farmed shrimp shells for arsenic removal because it is natural, low-cost, and environmentally friendly. Shrimp shells can be a good replacement for the current expensive methods to remove heavy metals from solution.

Friedel's salt was used by Zhang et al. [33] as a low-cost, efficient adsorbent. Approximately 92% As[V] could be removed at low pH and at a solid–liquid ratio of 0.2 g/L with an initial arsenic concentration of 2 mg/L. Cerium-loaded cation exchange resin has been used [34] on adsorption of arsenate and arsenite from aqueous solutions. The adsorption capacity of the resin is found to improve by impregnating cerium into the cation exchange resin.

Chitosan, a biopolymer extracted from the wastes of the seafood industries has also been used [35] as a cost-effective adsorbent for arsenic removal from contaminated water and it could remove over 90% of arsenic from contaminated water.

3.4 SYNTHESIS OF LOW-COST ADSORBENTS FOR ARSENIC

Activated alumina has been applied in many instances as an adsorbent to remove arsenic from contaminated water. When the alumina surface becomes sufficiently saturated with arsenic, it is necessary to regenerate the alumina. Adsorbed arsenic is removed from the alumina surface by contacting the saturated alumina with caustic soda. A subsequent neutralization with sulphuric acid is done for reusing alumina. But after two to three such regenerations, 20–30% adsorption capacity of alumina is lost. However, the main advantage of alumina is that its low cost and relatively high capacity for arsenic adsorption makes the process economical. Rate and degree of arsenic removal from water by alumina depends on the oxidation states of arsenic and pH of feed water. As(V) removal from water is more efficient compared to the removal of As(III). Use of chlorine or other oxidizing agents enhance arsenic adsorption. A pH in the range of 5.5–6.0 is found to be very effective in arsenic removal. Activated carbon and activated alumina have been the two most widely used adsorbents in arsenic removal. However, considering the cost of raw material and the cost of regeneration, such adsorbents can be made more attractive through cutting cost in the synthesis process. One such method of synthesis may be partial thermal dehydration, by which activated alumina can be synthesized at low cost.

This section deals with synthesis of activated alumina based on a method adopted by Sen and Pal [10]. The major discussions cover (1) preparation of an alumina-based adsorbent by partial thermal dehydration rather than by conventional gel precipitation method; (2) finding out the effects of the operating parameters on developing active surface area on the adsorbent for industrial scale up; (3) assessing whether the new method was cheaper than the gel precipitation method; (4) investigating whether the adsorbent is effective in removal of arsenic from a contaminated groundwater well up to the WHO-prescribed level; and (5) finding out the breakthrough curve of adsorption in a continuous column to determine the replacement requirement.

3.4.1 Partial thermal dehydration method

Since adsorbents need frequent replacements or regeneration when the bed gets exhausted, it is imperative to produce such adsorbents through low-cost methods so that the overall process remains economically viable. This section describes development of one such low-cost method. Activated alumina-based adsorbent with a high surface area is prepared following partial thermal dehydration of a gibbsite precursor. Effects of dehydration temperature, residence time, rate of increase of temperature, and particle size on the development of the active surface area of the adsorbent are described. The operating parameters have a profound impact on active surface area development. Brunauer–Emmett–Teller (BET) surface area (by nitrogen adsorption) and ignition losses are determined for all the samples. An adsorbent of surface area of around 335–340 m^2/g could be developed when dehydrated at 500 °C for a residence time of 30 min in a rapid heating system (rate of increase of temperature 200 °C/min) with a particle size of 200 mesh (85%). The arsenic adsorption capacity of this adsorbent is determined both in batch and column studies. The adsorbent is found to be very effective in removing arsenic. The adsorbent placed in the column could successfully remove arsenic from water up to a level below 10 ppb for more than 6,000 bed volume water.

A general observation [36] is that for large-scale treatment, the physico-chemical separation technique is possibly the best for the developing South East Asian countries. However, for small-scale treatment facilities like community water filters, an activated alumina-based adsorption technique can be an option provided it is produced at low cost sine adsorbent needs periodic replacement. Activated alumina-based adsorption columns can be very effective point-of-use water treatment devices by virtue of good mechanical properties, high specific surface area, and stability of activated alumina under most reaction conditions [37–41]. However considering the necessity of frequent replacement of the saturated adsorbent bed, the cost of synthesis of such an adsorbent must be kept low to make it affordable to the affected milieu across the world.

There are different methods for manufacturing active alumina adsorbents. It is produced mainly by gel precipitation of gibbsite powder with sodium hydroxide and sulfuric acid. This method gives more chemically pure activated alumina with well-defined physical parameters, which is used for different catalyst preparations. The cost of this active alumina is very high and, therefore, use of this adsorbent for water treatment is a costly affair in developing countries like Bangladesh and India, where vast rural areas have been affected by contamination of groundwater by arsenic. The alternate route for production of active alumina is by partial thermal dehydration of gibbsite powder. This process comprises a thermal treatment, at a suitable temperature for a suitable time, which removes 28–31% of water of crystallization from the gibbsite, leaving a loss on ignition of 4–7 wt%. This gives rise to a high active surface area, which is considered one of the controlling factors for arsenic adsorption.

During adsorbent development, parameters like dehydration temperature, rate of increase of temperature, residence time, and particle size are found to have significant effects on generation of specific surface areas of the adsorbent. The next few sections will help explain the critical preparation parameters and their combined effects on production of activated alumina in an environmentally benign way through a low-cost approach. So, an adsorption column is operated with this newly synthesized activated alumina powder (0.5–1.0 mm) to find its efficiency in removing arsenic from water.

3.4.1.1 Materials and procedures
The raw material used for the thermal dehydroxylation is known mineralogically as gibbsite and chemically as aluminum trihydroxide, $Al(OH)_3$. Commercial grade gibbsite powder as specified in Table 3.1 is used.

3.4.1.2 Equipment
Thermal hydroxylation of the gibbsite powder for slow heating and fast heating, respectively, is done in a muffle furnace, which works in the temperature range of 1200 °C ± 10 °C and in an electric oven in the temperature range of 0–700 °C. A standard Pot Mill made of stainless steel can be operated at 55 rpm for mechanical disintegration of raw gibbsite powder.

3.4.1.3 Measurement
The surface area of the sample is measured in a surface area analyzer. The instrument works on the theory first proposed by Brunauer–Emmett–Teller, known as the BET theory. According to this theory, at known partial

Table 3.1 Physicochemical Specification of Gibbsite Powder
Chemical Analysis

Fe_2O_3, % by wt	0.008
SiO_2, % by wt	0.005
Na_2O, % by wt	0.280
Al_2O_3, % by wt	65.00
H_2O, % by wt	Rest
Mesh size	
+50 mesh	30
+100 mesh fraction	80
+200 mesh fraction	100
Loss on ignition, %	34–36
Free moisture content	4%
Surface area, m^2/g	5

pressure, the amount of N_2 adsorbed by the sample at liquid N_2 temperature can be expressed as

$$(P/P_0) \text{ Vs } (1 - P/P_0) = (1/V_m C) + (C - 1)(P/P_0)/(V_m C) \qquad (3.9)$$

where P/P_0 = partial pressure of adsorbed gas, V_s = volume adsorbed at P/P_0 at 0 °C, V_m = monolayer volume, C = constant.

3.4.1.4 Synthesis procedure

For preparation of adsorbent with high surface area by partial thermal dehydration, a measured amount of dry gibbsite powder may be taken in a stainless steel bowl for heating in a muffle furnace at temperatures from 200 °C to 700 °C with a temperature interval of 50 °C. The rate of increase of temperature is kept at 10 °C/min and dehydration is done for 30 min. The bowl is immediately placed in desiccators to avoid moisture adsorption. The preparation scheme is shown in Figure 3.2.

To determine the effect of the time at the dehydration temperature, the same quantity of raw gibbsite powder is taken and dehydrated at 400, 450, and 500 °C (as the maximum surface area was obtained in this temperature zone) with time of dehydration 15, 30, and 45 min at the three temperatures. The samples are heated to 350, 400, 450, 500, and 550 °C in an electric oven with rate of increase of temperature of 200 °C/min to find out the effects of residence time and rate of heating. Samples are kept for a residence time of 15, 30, 45, 60, 75, 90, 105, and 120 min in the oven for each temperature.

To determine the effect of particle size for development of surface area, a measured amount of gibbsite powder is first dried for 6–8 hr in a tray dryer at 110 °C and after removing the free moisture the powder is milled for 1, 2, and 3 hr, respectively, in a standard Pot Mill. Sieve analysis as shown in

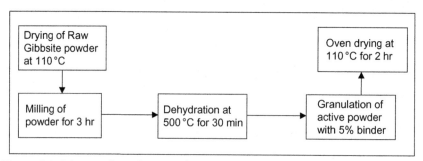

Figure 3.2 Scheme of production of activated alumina powder by partial thermal dehydration.

Table 3.2 Sieve Analysis of the Milled Gibbsite Powder

Milling Time (h)	Sieve Analysis	Surface Area (m²/g) at 500 °C
0	+50–30 +100–80 +200–100 +325–Nil	269
1	+50–Nil +100–65 +200–100 +325–Nil	316
2	+50–Nil +100–Nil +200–60 +325–100	327
3	+50–Nil +100–Nil +200–1% +325–85%	336

Table 3.2 is done for the milled powder. This milled powder is then dehydrated at 450, 500, and 550 °C in the electric oven for a residence time of 15, 30, and 45 min in each case.

The surface area and ignition loss are measured for all the samples to finally find out the combined effect of dehydration temperature, residence time, rate of heating, and particle size on the development of surface area. The activated alumina powder thus formed is granulated in a granulator with the help of spraying a 5% acetic acid binder. Granules are then air-dried and oven-dried to remove surface moisture. The process of the preparation of activated alumina granules is shown in Figure 3.2.

Granules from 0.50 to 0.9 mm are used for the determination of arsenic removal capacity.

Table 3.3 shows the physico-chemical specification of newly developed activated alumina granules.

3.4.1.5 Determination of arsenic adsorption capacity of the active alumina

For assessing adsorption capacity, normally 100 mL of 200 ppb As(III) and As(V) solution is taken in a 150-mL bottle, and agitated at 180 rpm for 16 hr at 30 °C. To study the effect of adsorbent dose on As(V) and As(III) removal, the adsorbent dose is varied from 5 to 50 g/L. The pH of the solution is

Table 3.3 Physico-chemical Specification of Activated Alumina Granules Produced by Partial Thermal Dehydration

Physical characteristics

Surface area, m^2/g	335
Pore volume, cm^3/g	0.5
Bulk density, g/cm^3	0.8 ± 0.05

Chemical composition, %

Al_2O_3	92.0
SiO_2	0.3
Fe_2O_3	0.05

maintained at 7 ± 0.1 by checking the pH value every 2 hr and adding N/10 HNO_3 dropwise.

To find out the effect of arsenic concentration of the raw water on the removal efficiency, a synthetic water sample containing 100–1000 ppb of arsenic (As(III) and As(V), separately) is added to 2.5 g (optimum dose) of adsorbent in 100 mL water. The sample is agitated for 16 hr in a shaker at 30 °C at 180 rpm. The pH of the solution is measured after each 2 hr and adjusted to 7 ± 0.1 by addition of N/10 HNO_3. Finally concentration of residual arsenic in solution is measured using the atomic absorption spectrophotometric method.

3.4.1.6 Arsenic adsorption column

A typical apparatus that may be employed for adsorption of arsenic in a packed column is shown in Figure 3.3. An adsorption column may be directly connected to a hand tube well pumping arsenic-contaminated groundwater as shown in Figure 3.4. Virgin-activated alumina granules of size 0.5–0.9 mm may be packed in a column made of glass, polycarbonate material, or stainless steel.

Arsenic-contaminated groundwater or water spiked with arsenic solution may be used to find the column performance of the newly formed activated alumina. The raw water o passes down from the top of the column to the bottom at a specified rate based on column dimension. Initially all arsenic ions are adsorbed by the media resulting in zero effluent concentration. Most of the solute is removed initially by a narrow band of activated alumina granules at the top of the column (adsorption zone). As operation continues the adsorption zone proceeds toward lower layers of activated alumina. Ultimately when the adsorption zone reaches the bottom of the column, arsenic concentration in the outlet begins to increase. A plot of solute

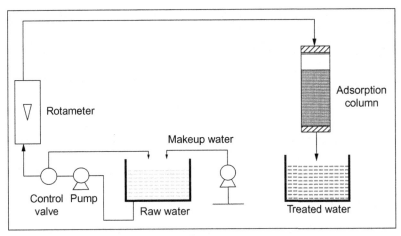

Figure 3.3 Typical arsenic adsorption column.

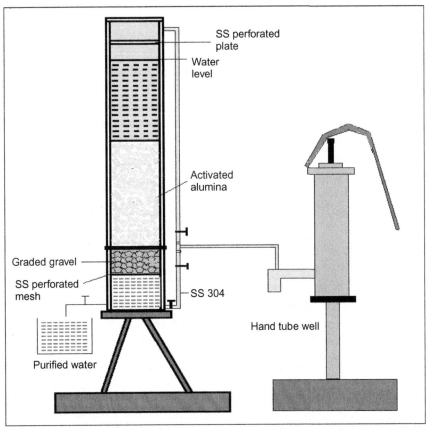

Figure 3.4 System of arsenic adsorption column directly connected to hand tube well in arsenic-affected area.

concentration in the effluent (mg/L) versus bed volume referred to as the "breakthrough curve" is obtained by plotting arsenic concentration in the effluent against the bed volume.

3.4.1.7 Analytical procedures

Analysis of the samples for arsenic concentration may be done following the flame-FIAS technique in an atomic adsorption spectrophotometer. In the flame-FIAS technique arsenic is analyzed after its conversion to volatile hydride, which is formed through the following reactions:

$$NaBH_4 + 3H_2O + HCl = H_3BO_3 + NaCl + 8H$$
$$As_3 + H\,(excess) \longrightarrow AsH_3 + H_2\,(excess)$$

Volatile hydride is then transported to the quartz cell of the atomic adsorption spectrophotometer, where it is converted to gaseous arsenic metal atoms at 900 °C in air acetylene flame and analysis is done at 193.7 nm wavelength using a hollow cathode lamp. Samples to be analyzed for arsenic are prereduced using a reducing solution consisting of 5% (w/v) potassium iodide (KI) and 5% (w/v) ascorbic acid. The reduced sample is allowed to stand at room temp for 30 min. In hydride generation, 0.2% sodium borohydrate (NaBH$_4$) in 0.05% NaOH was used. Ten percent (v/v) HCl is used as carrier solution. Hydride generator is purged using 99.99% pure nitrogen. pH of the oxidation unit and the filtered water is measured by a pH-ion meter. Percentage removal of arsenic is calculated using the initial value (C_{As0}) and the residual value (C_{As}) of arsenic concentration in feed water and treated water (permeate), respectively:

$$\% \text{ removal of arsenic} = (1 - C_{As}/C_{As0}) \qquad (3.10)$$

3.4.2 Effects of operating conditions on characteristics of the developed adsorbent

3.4.2.1 Effect of temperature on surface area development

The surface area of raw gibbsite powder is approx 5 m^2/g, which does not show any significant increase in the surface area up to a temperature of 200 °C but thereafter a rapid increase in surface area with rapid release of structural water and a sharp peak is observed in the temperature range of 400–500 °C when heated for 30 min in a slow heating system (rate of increase of temperature 10 °C/min). Figure 3.5 shows the variation in surface area when heated at various temperatures and a maximum surface area of 280 m^2/g is achieved at 450 °C. It indicates that surface area increases with the rise in temperature as gibbsite gets converted to bohemite during

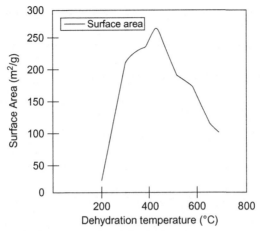

Figure 3.5 Effect of change in the dehydration temperature on surface area. Dehydration over 100–700 °C in a muffle furnace at heating rate of 10 °C/min.

heating. It is due to the fact that gibbsite remains in crystalline form at normal temperature but it converts to amorphous alumina due to heating to a range of 450–500 °C, which is having more surface area than crystalline alumina.

3.4.2.2 Effect of residence time on surface area development

Figure 3.6 exhibits the effect of residence time on the generation of surface area and loss on ignition. The residence time at a particular dehydration

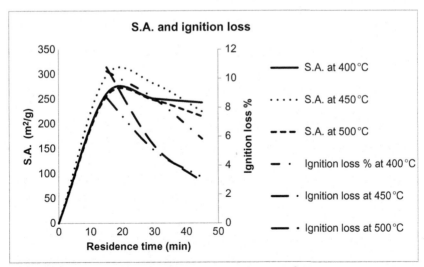

Figure 3.6 Effect of residence time (min) on surface area (m²/g) and ignition loss (%) of gibbsite powder at 400, 450, and 500 °C in a slow heating system.

temperature is varied for determination of the optimum condition of thermal dehydration to get a high surface area active material directly from partial dehydration of gibbsite powder. It is found that a residence time of 15 min at a temperature of 450 °C is the most suitable residence time for development of high surface area when heating is done in a slow heating system. From the figures it is observed that the surface area development at a particular temperature depends on heating period. The surface area of the order of 295–300 m^2/g can be developed with an ignition loss of 8–9%.

Thus, by reducing the time of stay, larger surface area can be developed. Surface area formation depends on the presence of 0.5 H_2O as shown by Blanchin [42] in the alumina. It is a matter of controls on the heating system, how the water should be released at a particular temperature. Staying for less or more time will have an effect on release of water molecule from the alumina powder and will have effect on surface area development. Maximum surface area is developed when the water molecule as above remains attached with alumina. Staying at a particular temperature reduces the presence of ½ molecule of H_2O, thereby reducing the surface area.

3.4.2.3 Effect of rapid dehydration

It is observed that the optimum dehydration temperature shifts from 450 °C to 500 °C when the same test is carried out under very rapid dehydration. The surface area increases to a value of 310 m^2/g at 500 °C and with a residence time of 30 min as exhibited in Figure 3.7. The reason behind the high surface area is probably the development of amorphous ρ,χ, η, and γ alumina, which were obtained by rapid heating and rapid removal of water vapor as reported by Whittington [43].

3.4.2.4 Effect of particle size

Surface area of the thermally dehydrated gibbsite may be increased by applying mechanical grinding for reduction of particle size.

Figure 3.8 reveals that no significant increase in the surface area is obtained after 1 hr milling, but after 2 and 3 hr (Figures 3.9 and 3.10, respectively), milling surface area increases by about 5% and 8.5%, respectively, over the unground material dehydrated under the same operating conditions. The particle size of the raw material has an effect on the surface area of an activated alumina powder when derived from partial dehydration of gibbsite. The sieve analysis shows the size range of the particle. This finely ground particle when partially dehydrated at 500 °C for 30 min gives a

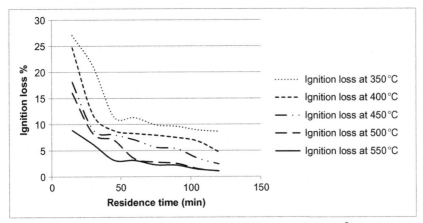

Figure 3.7 Effect of rapid dehydration on (a) development of S.A. (m^2/g); (b) ignition loss % temperature: 350, 400, 450, 500, and 550 °C, rate of increase of temp 200 °C/min, residence time (min), surface area (m^2/gm), and ignition loss %.

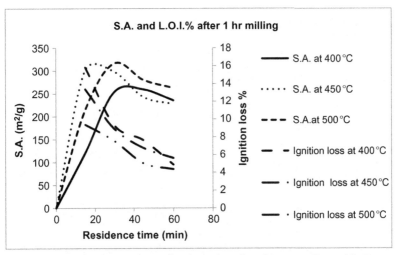

Figure 3.8 Effect of particle size on development of surface area. Oven-dried material milled for 1 hr dehydrated at 400, 450, and 500 °C.

maximum surface area of 336 m^2/g. This is due to the fact that in case of smaller particle size, the water vapor formed can rapidly diffuse out of the trihydroxide particle. The smaller particle sizes thus resist the buildup of vapor pressure inside the particle, thereby prohibiting formation of

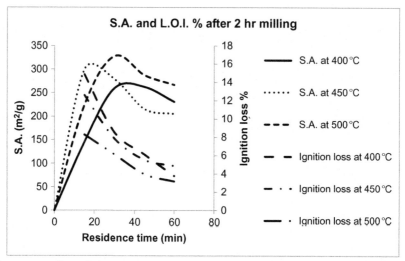

Figure 3.9 Effect of particle size on development of surface area. Oven-dried material milled for 2 hr dehydrated at 400, 450, and 500 °C.

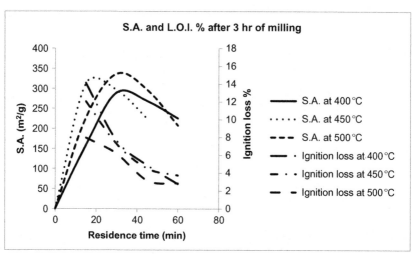

Figure 3.10 Effect of particle size on development of surface area. Oven-dried material milled for 3 hr dehydrated at 400, 450, and 500 °C. S.A. = surface area; L.O.I. = loss of ignition.

bohemite, which has a smaller surface area. Literature shows that relatively less bohemite is formed due to the small size of the crystal. Thus the best product is obtained when dehydrated in a fast heating system at 500 °C for 30 min after milling for 3 hr. The surface area thus formed was 336 m^2/g.

3.4.2.5 Efficiency in removing arsenic from water

Figure 3.11 shows the effect of adsorbent dose on percentage removal of arsenic. Percentage removal is found to increase with adsorbent dose initially until it reaches around 25 g/L. Beyond this value, however, no further increase in percentage removal accompanies the increase in dose. Thus the optimum adsorbent dose for contaminated water with 200–400 μg/L As concentration is found to be around 25 g/L.

In the batch test, the adsorbent removes as high as 90% As(V) and 42% As(III) for feed concentration up to 400 ppb. This is because the anionic As(V) is more attracted by activated alumina than neutral As(III). Above 400 ppb, removal percentage decreases gradually, resulting in the lowest adsorption of 72% and 35% for As(V) and As(III), respectively, as exhibited in Figure 3.12 when arsenic concentration in raw water is 1000 ppb.

Figure 3.13 shows a breakthrough curve. The point on the breakthrough curve where the arsenic concentration in the effluent turns 0.01 mg/L is the breakthrough point. The point where arsenic concentration reaches 90% of the influent (900 ppb) is the point of exhaustion. The accumulation and subsequently the removal of arsenic are mainly dependent on the quantity of the adsorbent available in the bed.

Figure 3.13 indicates that during continuous run on the adsorption column, breakthrough bed volume is 6,000, which means effluent arsenic concentration up to this bed volume has a value below 10 μg/L. The exhaustion point corresponds to a bed volume of 13,000, when the effluent arsenic

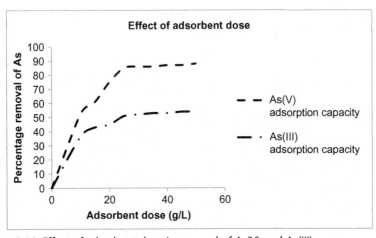

Figure 3.11 Effect of adsorbent dose in removal of As(V) and As(III).

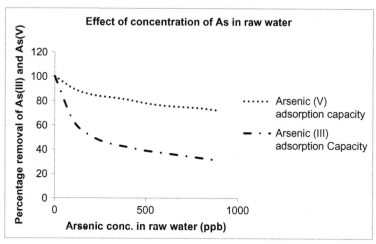

Figure 3.12 Effect of concentration of As(V) and As(III) in raw water.

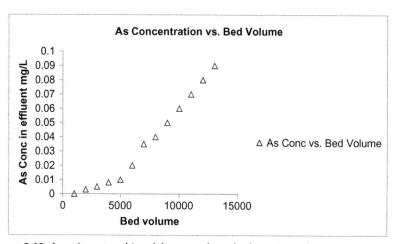

Figure 3.13 Arsenic removal in a laboratory-based adsorption column.

concentration is found to be 90% of the influent arsenic concentration. Thus the activated alumina produced by partial thermal dehydration of gibbsite powder is found to efficiently remove arsenic from drinking water.

It is observed that a residence time of 15 min gives better results than 30 min when heated at 450 °C in a slow heating system. Dehydration temperature shifts from 450 to 500 °C when heated in a rapid dehydration system instead of a slow heating system. The surface area improves by about

9% when particle size changes from 100% +200 mesh to 1% +200 mesh. The maximum surface area in this case is 336–340 m^2/g when dehydrated at 500 °C in a rapid dehydration system for 30 min and after milling for over 3 hr. The ignition loss (LOI) is determined in all the cases as it gives an idea of the amount of structural water remaining in the active alumina powder. With the increase of dehydration temperature, the ignition loss decreases, keeping much less water in it, thereby reducing the surface area. The adsorbent thus prepared by this active alumina powder shows As(V) removal over 90% whereas As(III) removal is only 42% for raw water having 400 ppb arsenic. Neutral characteristics of As(III) and anionic characteristics of As(V), respectively, may explain such a difference in adsorption. However, the column performance in removing as(V) is found to be satisfactory where the arsenic level in water can be reduced to well below 10 ppb for more than 6,000 bed volumes. The exhaustion bed volume is found to be more than 13,000.

3.4.3 Modeling and simulation of column adsorption for scale up

3.4.3.1 Kinetics of adsorption and adsorption isotherm

Adsorption isotherms are used for understanding the mechanism and quantifying the distribution of the adsorbate between the liquid phase and solid adsorbent phase at equilibrium during the adsorption process. For adsorption modeling there are several empirical models; among them the most commonly used in sorption studies are the Langmuir, Freundlich, Redlich–Peterson, and BET [1–4] models. The Langmuir model is valid for monolayer adsorption on a surface with a finite number of adsorption sites of equal energy and is expressed as

$$A_e = \frac{A_{m,e} k_L C_e}{1 + k_L C_e} \tag{3.11}$$

where A_e and A_m, respectively, represent the amount of adsorption and the maximum amount of adsorption per unit of adsorbent used at equilibrium arsenic concentration, C_e. k_L is the equilibrium coefficient of the Langmuir model. The linearized form of this model is expressed as

$$\frac{C_e}{A_e} = \frac{1}{k_L A_{m,e}} + \frac{C_e}{A_{m,e}} \tag{3.12}$$

The Freundlich model is an empirical equation developed based on the assumption that the adsorbent has a heterogeneous surface composed of different classes of adsorption sites, and each site can be modeled by the following equation:

$$A_e = k_F C_e^{1/n} \tag{3.13}$$

where k_F is the equilibrium coefficient of the Freundlich isotherm and n represents the intensity constants of the adsorbents. The linearized form this model is

$$\ln A_e = \ln k_F + \frac{1}{n} \ln C_e \tag{3.14}$$

The Redlich–Peterson model combines models of Freundlich and Langmuir and also describes the heterogeneity of the sorbent surface and a certain number of adsorption sites with the same adsorption potential. The equation by Redlich–Peterson is the following:

$$A_e = \frac{k_R C_e}{1 + a C_e^b} \tag{3.15}$$

where k_R is the equilibrium coefficient of the Redlich–Peterson isotherm and a, b represent as an equation constant. The linearized form this model is

$$\ln \left(\frac{k_R C_e}{A_e} - 1 \right) = \ln a + b \ln C_e \tag{3.16}$$

The BET model provides coverage of the surface with adsorbate multilayers, and as active sites have different energies, multilayers can be formed in different parts of the surface and the BET equation can be expressed as

$$A_e = \frac{A_m k_{BET} C_e C_{Fl}}{(C_{Fl} - C_e) + (C_{Fl} + k_{BET} C_e - C_e)} \tag{3.17}$$

where k_{BET} is the equilibrium coefficient of BET isotherm and C_{Fl} is the filled layer concentration. The linearized model of the BET isotherm is

$$\frac{C_e}{A_e(C_{Fl} - C_e)} = \frac{1}{A_m k_{BET}} + \frac{k_{BET} C_e - C_e}{A_m k_{BET} C_{Fl}} \tag{3.18}$$

The equilibrium constants (k_L, k_F, k_R, and k_{BET}) and different constants (a, b, and n) of the equations are calculated from the slope and the intersection of the line graph by plotting $\frac{C_e}{A_e}$ vs. C_e (Eq. 3.2), $\ln A_e$ vs. $\ln C_e$ (Eq. 3.4), $\ln \left(\frac{k_R C_e}{A_e} - 1 \right)$ vs. $\ln C_e$ (Eq. 3.6), and $\frac{C_e}{A_e(C_e - C_{Fl})}$ vs. $\frac{C_e}{C_{Fl}}$ (Eq. 3.18).

3.4.4 Dynamic modeling of adsorption in a fixed-bed column

Dynamic mathematical model can be developed for predicting break-through curves for fixed bed adsorption columns with the following assumptions:

- Single component, isotherm system with plug flow axial dispersion
- Liquid density is constant throughout the fixed bed adsorption system
- The adsorption granules are spherical in shape
- The iron-pellet mass transfer is due to diffusion of adsorbate molecules into the adsorbent pores

The purpose of fixed-bed column adsorption is to reduce the concentration of the contaminant in the effluent below the permissible limit. The breakthrough curve of the system predicts the lifetime of the bed and regeneration time. The shape of the breakthrough curve depends on the inlet flow rates, concentration, column diameter, bed height, and the adsorbent's physico-chemical properties. The fixed bed of certain dimensions has a definite capacity to adsorb the solute entering the bed, which is equivalent to that of adsorption of pollutant not only until the time of achieving equilibrium but also upon the transfer mechanism and rate of adsorption. Considering a spherical adsorption pellet, the material balance equation for diffusion path length z may be written as

$$-D_L\frac{\partial^2 C}{\partial Z^2} + \frac{\partial(VC)}{\partial Z} + \frac{\partial C}{\partial t} + \left(\frac{1-\varepsilon}{\varepsilon}\right)\rho_s\frac{\partial\rho}{\partial t} = 0 \qquad (3.19)$$

Rewriting the equation with regard to $\partial C/\partial t$, we get

$$\frac{\partial C}{\partial t} = D_L\frac{\partial^2 C}{\partial Z^2} - C\frac{\partial V}{\partial Z} - V\frac{\partial C}{\partial Z} - \left(\frac{1-\varepsilon}{\varepsilon}\right)\rho_s\frac{\partial\rho}{\partial t} \qquad (3.20)$$

where D_L = axial dispersion coefficient. When the fluid flows through the packed bed there is a tendency for axial mixing to occur. Any such mixing is undesirable because it decreases the separation efficiency. So axial dispersion is minimized to get better separation efficiency. In this model the effects of all mechanisms that contribute to axial mixing are lumped together into a single axial dispersion coefficient.

There are two main mechanisms that contribute to axial dispersion, molecular diffusion and turbulent mixing arising from the splitting and recommendation of flows around the adsorbent particles. Considering these two effects the dispersion coefficient may be represented as follows:

$$D_L = \gamma_1 D + \left(\gamma_2 r_p 2V\right) \qquad (3.21)$$

where following Wicke we get

$$\gamma_1 = 0.45 + 0.55\varepsilon \qquad (3.22)$$

and

$$\gamma_2 = \frac{\gamma_1 \varepsilon}{R_e S_c} + \frac{1}{3.35 r_p \left(1 + \dfrac{\beta \gamma 1 \varepsilon}{R_e S_c}\right)} \qquad (3.23)$$

where $r_p < 0.15$ cm and $\beta = 13.0$. This is of considerable practical importance since it is evident that the advantage of reduced pore diffusion resistance that is gained by reduction of particle size can easily be offset by the increased axial dispersion once the particle diameter is reduced below 0.3 cm. The results here are for an average particle diameter of 0.07 cm.

The initial and boundary conditions are

$$C = C_0 \qquad\qquad \text{at } Z = 0, t = 0$$

$$\frac{\partial C}{\partial Z} = 0 \qquad\qquad \text{at } Z = L, t \geq 0$$

$$D_L \frac{\partial C}{\partial Z} = -V_0(C_0 - C) \quad \text{at } Z = 0, t > 0$$

Due to adsorption in the packed bed the superficial velocity is not constant over the bed height. This is due to the adsorption of arsenic into adsorption pores. Change in the velocity along the height of the bed may be calculated by the following equation:

$$\rho_s \frac{\partial q}{\partial t}(1 - \varepsilon) = -\rho_1 \frac{\partial V}{\partial Z} \qquad (3.24)$$

Boundary conditions are

$$V = V_0 \quad \text{at } z = 0, t > 0$$

$$\frac{\partial V}{\partial Z} = 0 \quad \text{at } z = L, t > 0$$

The transport of the absorbable species from the bulk of the liquid to phase to the external surface of adsorbent pellets constitutes an important step in the overall uptake process.

For single species adsorption with spherical pellets, the interphase mass transfer rate may be expressed as

$$\frac{\partial q}{\partial t} = \frac{3k_f}{a_p \rho_p}(C - C_s) \qquad (3.25)$$

where k_f is the mass transfer coefficient; k_f is found using the correlation of Wilson and Geankoplis.

$$Sh = \frac{2k_f r_p}{D} = \left(\frac{1.09}{\varepsilon}\right) R_e^{0.33} S_c^{0.33}$$

This will be effective for $0.0015 < R_e < 55$.

Mass transfer within the adsorbent pellet (assuming pore diffusion is controlling) is

$$\varepsilon_p \frac{\partial C}{\partial t} + \rho_p \frac{\partial q}{\partial t} = \left(\frac{D_p}{r^2}\right) \frac{\partial}{\partial r}\left[r^2 \frac{\partial C}{\partial r}\right] \qquad (3.26)$$

Initial and boundary conditions are

$$C = 0, q = 0 \qquad \qquad \text{at } 0 < r < r_p \text{ and } t < 0$$

$$\frac{\partial C}{\partial r} = 0 \qquad \qquad \text{at } r = 0, t > 0$$

$$k_f(C - C_s) = D_p \frac{\partial C}{\partial r} \qquad \text{at } r = r_p$$

These equations were solved numerically by the backward implicit method to find out the bulk concentration and the concentration in pore at various radial and axial positions.

The model equations can be solved numerically using the backward implicit method to obtain the following equation.

$$C_{b\,z-1,\,t+1}\left[(D_L \Delta t / \Delta z^2) + \{V_z \Delta t / (2\Delta Z)\}\right]$$
$$- C_{b\,z,\,t+1}\left[1 + (2D_L \Delta t / \Delta z^2) + \{\Delta t (V_z - V_{z-1})/\Delta Z\}\right]$$
$$+ \{3(1-\varepsilon) \cdot k_f \Delta t / (\varepsilon \cdot a_p)\}] + C_{b\,z+1,\,t+1}\left[(D_L \Delta t / \Delta z^2) + \{V_z \Delta t / (2\Delta Z)\}\right]$$
$$= -\left[3(1-\varepsilon) \cdot k_f \Delta t\, C_{s\,z,t} / (\varepsilon \cdot a_p)\right] - C_{b\,z,t} \qquad (3.27)$$

With Boundary conditions:

(1)   At $t = 0$, $z = 0$

$$C_b = C_{b,\,in}$$

(2)   At $t > 0$, $z = 0$

$$D_L \Delta t / \Delta z^2 [1 + (V_0 \Delta z / D_L)] C_{b\,z,\,t+1} - C_{b\,z+1,\,t+1} = (V_0 \Delta z / D_L)] C_{b,\,in}$$

(3)   At $t >= 0$, $z = L$

$$C_{b\,z,\,t+1} - C_{b\,z-1,\,t+1} = 0$$

The axial velocity gradient of Eq. (3.18) was discretized into finite difference form as

$$V_{Z,t} = V_{Z-1,t} - \left[3(1-\varepsilon) \cdot k_f \Delta z / (\rho_1 \cdot a_p)\right] (C_{bz,t} - C_{sz,t}) \qquad (3.28)$$

With Boundary conditions:
(1)   At $t > 0$, $z = 0$
$$V_{Z,t} = V_0$$
(2)   At $t > 0$, $z = L$
$$V_{Z,t} - V_{Z-1,t} = 0$$

And mass transfer within the porous absorbent particle (Eq. 3.20) was discretized to give

$$C_{r-1,t+1}\left[\left(D_p/\Delta r^2\right) - \left\{D_p/2(r-1)\Delta r\right)\right\}]$$
$$- C_{r,t+1}\left[\left(\varepsilon_p/\Delta t\right) + \left(2D_p/\Delta r^2\right)\right] + C_{r+1,t+1}\left[\left(D_p/\Delta r^2\right) + \left\{D_p/2(r-1)\Delta r)\right\}\right]$$
$$= \left(\varepsilon_p/\Delta t\right) C_{r,t} + \left\{(1-\varepsilon_p)^P/\Delta t\right\}(q_{r,t+1} - q_{r,t})$$

$$(3.29)$$

With Boundary conditions:
(1)   At $t = 0$, $0 < r < r_p$
$$C_{r,t} = q_{r,t} = 0$$
(2)   At $t > 0$, $r = 0$
$$C_{r,t+1} - C_{r+1,t+1} = 0$$
(3)   At $t > 0$, $r = r_p$
$$\left[1 + \left(k_f \Delta r / D_p\right)\right] C_{r+1,t+1} - C_{r,t+1} = \left(k_f \cdot \Delta r / D_p\right) C_{bz,t+1}$$

Three major parameters used in the model equations are liquid dispersion coefficient (DL), pore diffusivity (D_p), and mass transfer coefficient (K_f). Initial values of these parameters are obtained from available correlations of Perry and Tien [44]. The values for various column parameters from the model are shown in Table 3.4.

3.4.4.1 Breakthrough studies for model validation

C_f/C_0 against time (hr) represents a theoretical breakthrough curve. The experimental breakthrough curve at 10-cm bed depth (initial As conc. 1 mg/L) at a flow rate of 4 mL/min is compared here with the theoretical breakthrough curve developed using a pore diffusion model. The shape of the curves shows that both internal and external resistances are significant [45]. It is shown that at 10 cm the bed depth theoretically predicted value almost matches the experimental value. Therefore the model can be used

Table 3.4 Various Column Parameters that Can Be Obtained from the Model

Parameters	Value
Bed height, m	0.2
Inlet concentration, kg/m^3	1×10^{-3}
Final outlet concentration, kg/m^3	9×10^{-3}
Reynolds number	0.212
Schmidt number	3.85×10^9
Peclet no.	8.15×10^8
Flow rate, m^3/s	0.066×10^{-6}
Initial superficial velocity, m/s	2.12×10^{-4}
Pore diffusion coefficient, D_p, m^2/s	2.6×10^{-16}
Axial dispersion coefficient, D_L, m^2/s	6.328×10^{-10}
Mass transfer coefficient, K_f m/s	3.76×10^{-5}

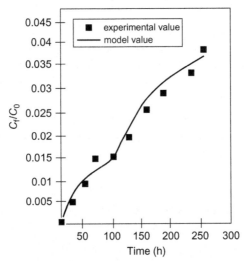

Figure 3.14 Comparison of experimental breakthrough curve with theoretical ($C_0 - 1$ mg/L, flow rate 4 mL/min); (---) theoretical; (♦) experimental.

to predict the breakthrough curve for arsenic adsorption in a fixed column on newly developed adsorbent. The result is shown in Figure 3.14.

3.4.4.2 Error Analysis

How well experimental data are correlated with the model-predicted values can be calculated as per the following formulae.

1. Root mean square error (RMSE) has been calculated by the following equation:

$$RMSE = \sqrt{\frac{\sum_{i=1}^{n} (P_i - E_i)^2}{n}}$$

where

E_i = Experimental value
P_i = Predicted (model) value
n = number of points analyzed

2. Relative error (RE) was calculated as per the following formula:

$$RE = \frac{RMSE}{\bar{E}}$$

where $\bar{E}$ = *mean of experimental data.*

3. The Willmott Index of arrangement (d) has been calculated as per the following equation formula.

From the combination of the values of d and RE the model performance may be predicted as per the following relations [46]:

Combination of values	Model performance
$d \geq 0.95$ and $RE \leq 0.10$	Very good
$d \geq 0.95$ and $0.15 \geq RE > 0.10$	Good
$d \geq 0.95$ and $0.20 \geq RE > 0.15$	Acceptable
$d \geq 0.95$ and $0.25 \geq RE > 0.20$	Poor

3.5 TECHNOECONOMIC FEASIBILITY ANALYSIS

The cost factors associated with production of activated alumina by the gel precipitation method and by partial thermal dehydration of gibbsite powder are presented in Table 3.5. The data are collected from Oxide (India) Catalysts Pvt. Ltd. Table 3.5 shows that the cost of manufacturing alumina-based adsorbent with the present technique is substantially less than the cost associated with the gel precipitation method. The production process using partial thermal dehydration involves a much simpler scheme than a conventional gel precipitation process.

For removal of arsenic from contaminated groundwater a low-cost adsorbent based on activated alumina can be prepared by partial thermal dehydration of gibbsite. The adsorbent can successfully reduce the arsenic level in contaminated water well below the WHO-prescribed level

Table 3.5 Cost Comparison of Activated Alumina Produced by Gel Precipitation
Process and Thermal Dehydration Process for 1 MT Material*

Components	Gel precipitation method	Cost ($)	Thermal dehydration method	Cost ($)
Raw material	Aluminium sulfate 1560 kg @ $ 0.34/kg	530	Gibbsite powder 1500 kg @ $0.34/kg	510
	Gibbsite powder 1120 kg @ $0.34 /kg	381		
	Caustic soda 560 kg @ $0.68/kg	380	Binder	70
	Binder	40		
Utilities	Electricity & fuel	336	Electricity & fuel	432
Overhead & labor		212		140
Total		1879		1152

*Cost computation is based on Indian market price where 1$ = 60 Indian Rupees (INR) of 2014.

(10 µg/L). This simple scheme involves fewer steps, less energy, and little harsh chemicals and is not only cheaper than conventional gel precipitation methods but is environmentally benign also. Since most of the arsenic-affected people live in the rural backwaters of South East Asian countries where community water filters fail because of the need for frequent replacement of costly adsorbent material, the proposed manufacturing scheme is likely to be a highly promising one. In general, an adsorption column can simply be connected to a hand tube well for purifying arsenic-contaminated water. Initial capital investment in such point-of-use facilities is quite affordable. For treating 1 m³ of contaminated water by an alumina-based adsorbent, annualized operating cost is around 1 $ in Indian context, making it quite competitive with other conventional techniques like alum precipitation or iron precipitation. However, very often such facilities fail due to a lack of maintenance, which needs proper monitoring of the water quality being produced with time.

During the initial phase after installation of a treatment facility, the quality of water remains very good. But eventually the adsorbent bed gets completely saturated, rendering further treatment impossible. Commonly in such cases people are quite likely to be misled as to the quality of water in absence of a suitable provision for analysis of the treated water. Replacement of the adsorption bed needs to be done the moment it gets exhausted. Another technical problem that is often encountered in such adsorption column operation is clogging of the adsorption bed in the column by deposited iron and iron-based compounds. In most cases, arsenic-contaminated groundwater accompanies iron in high concentration. Precipitation and settling of such iron frequently clogs the bed material. Disposal of the spent adsorption material may be another problem area.

NOMENCLATURE

a_p specific surface area, m^2/g
r_p pellet radius, cm
C concentration of arsenic in water, mg/L
C_0 concentration of arsenic in the inlet water, mg/L
L height of the packed bed, cm
D_L axial dispersion coefficient, m^2/s
D_p pore diffusivity
V interstitial velocity, m/s
V_0 value of V at inlet
ρ_s density of the solid adsorbent, kg/m^3
ρ_l density of the liquid, kg/m^3
q amount of arsenic adsorbed per unit of activated alumina (mg/g)
ε bed porosity
C_b adsorbate concentration in bulk
C_s adsorbate concentration at the solid–pellet interface
K_f mass transfer coefficient

REFERENCES

[1] Langmuir I. The constitution and fundamental properties of solids and liquids. J Am Chem Soc 1916;38:2221.
[2] Brunauer S, Emmett PH, Teller E. Adsorption of gases in multimolecular layers. J Am Chem Soc 1940;62:1723.
[3] Freundlich HMF. Over the adsorption in solution. J Phys Chem 1906;57:385.
[4] Redlich O, Peterson DL. A useful adsorption isotherm. J Phys Chem 1959;63:1024.
[5] Manju GN, Raji C, Anirudhan TS. Evaluation of coconut husk carbon for the removal of arsenic from water. Water Res 1998;32(10):3062–70.
[6] Dobrowolski R, Otto M. Preparation and evaluation of Ni-loaded activated carbon for enrichment of arsenic for analytical and environmental purposes. Micropor Mesopor Mater 2013;179:1–9.
[7] Li W-G, Gong X-J, Wang K, Zhang X-R, Fan W-B. Adsorption characteristics of arsenic from micro-polluted waterby an innovative coal-based mesoporous activated carbon. Bioresour Technol 2014, http://dx.doi.org/10.1016/j.biortech.2014.02.069.
[8] Folkers K, Shovel J. U.S. Patent No. 2573 702; 1901.
[9] Barnitt JB. U.S. Patent No.1 868 869; July 26, 1932.

[10] Sen M, Pal P. Treatment of arsenic-contaminated groundwater by a low cost activated alumina adsorbent prepared by partial thermal dehydration. Desalin Water Treat 2009;11:275–82.

[11] Chutia P, Kato S, Kojima T, Satokawa S. Arsenic adsorption from aqueous solution on synthetic zeolites. J Hazard Mater 2009;162:440–7.

[12] Payne KB, Abdel-Fattah TM. Adsorption of arsenate and arsenite by iron-treated activated carbon and zeolites: effects of pH, temperature and ionic strength. J Environ Sci Health A Tox Hazard Subst Environ Eng 2005;40:723–49.

[13] Chutia P, Kato S, Kojima T, Satokawa S. Arsenic adsorption from aqueous solution on synthetic zeolites. J Hazard Mater 2009;162:440–7.

[14] Altundogan HS, Altundogan S, Tuèmen F, Bildik M. Arsenic removal from aqueous solutions by adsorption on red mud. Waste Manag 2000;20:761–7.

[15] Brunori C, Cremisini C, Massanisso P, Pinto V, Torricelli L. Reuse of a treated red mud bauxite waste: studies on environmental compatibility. J Hazard Mater 2005;117 (1):55–63.

[16] Li Y, Wang J, Luan Z, Liang Z. Arsenic removal from aqueous solution using ferrous based red mud sludge. J Hazard Mater 2010;177:131–7.

[17] Amin MN, Kaneco, Kitagawa T, Begum A, Katsumata H, Suzuki T, et al. Removal of arsenic in aqueous solutions by adsorption onto waste rice husk. Ind Eng Chem Res 2006;45:8105–10.

[18] Diamadopoulos E, Loannidis S, Sakellaropoulos GP. As(V) removal from aqueous solutions by fly ash. Water Res 1993;27(12):1773–7.

[19] Singh DB, Prasad G, Rupainwar DC. Adsorption technique for the treatment of As(V)-rich effluents. Colloids Surf A Physicochem Eng Asp 1996;111 (1–2):49–56.

[20] Bajpai S, Chaudhuri M. Removal of arsenic from ground water by manganese dioxide-coated sand. J Environ Eng 1999;125:782–7.

[21] Hanson A, Bates J, Heil D. Removal of arsenic from ground water by manganese dioxide-coated sand. J Environ Eng 2000;126:1160–1.

[22] Nguyen TV, Vigneswaran S, Ngo HH, Pokhrel D, Viraraghavan T. Specific treatment technologies for removing arsenic from water. Eng Life Sci 2006;6(1):86–90.

[23] Goswami A, Raul PK, Purkait MK. Arsenic adsorption using copper (II) oxide nanoparticles. Chem Eng Res Des 2012;90:1387–96.

[24] Olyaie E, Banejad H, Afkhami A, Rahmani A, Khodaveisi J. Development of a cost-effective technique to remove the arsenic contamination from aqueous solutions by calcium peroxide nanoparticles. Sep Purif Technol 2012;95:10–5.

[25] Maji SK, Pal A, Pal T. Arsenic removal from real-life groundwater by adsorption on laterite soil. J Hazard Mater 2008;151:811–20.

[26] Kundua S, Kavalakatt SS, Pal A, Ghosh SK, Mandal M, Pal T. Removal of arsenic using hardened paste of Portland cement: batch adsorption and column study. Water Res 2004;38:3780–90.

[27] Banerjee K, Amy GL, Prevost M, Nour S, Jekel M, Gallagher PM, et al. Kinetic and thermodynamic aspects of adsorption of arsenic onto granular ferric hydroxide [GFH]. Water Res 2008;42:3371–8.

[28] Li Y, Zhang FS, Xiu FR. Arsenic (V) removal from aqueous system using adsorbent developed from a high iron-containing fly ash. Sci Total Environ 2009;407:5780–6.

[29] Nguyen TV, Vigneswaran S, Ngo HH, Kandasamy J. Arsenic removal by iron oxide coated sponge: experimental performance and mathematical models. J Hazard Mater 2010;182:723–9.

[30] Lunge S, Singh S, Sinha A. Magnetic iron oxide (Fe_3O_4) nanoparticles from tea waste for arsenic removal. J Magn Magn Mater 2014;356:21–31.

[31] Tiana Y, Wua M, Linb X, Huangb P, Huang Y. Synthesis of magnetic wheat straw for arsenic adsorption. J Hazard Mater 2011;193:10–6.

[32] Chioa CP, Linb MC, Liaoa CM. Low-cost farmed shrimp shells could remove arsenic from solutions kinetically. J Hazard Mater 2009;171:859–64.

[33] Zhang D, Jia Y, Mab J, Li Z. Removal of arsenic from water by Friedel's salt (FS: $3CaO \cdot Al_2O_3 \cdot CaCl_2 \cdot 10H_2O$). J Hazard Mater 2011;195:398–404.

[34] Zongliang H, Senlin T, Ping N. Adsorption of arsenate and arsenite from aqueous solutions by cerium-loaded cation exchange resin. J Rare Earth 2012;30(6):563.

[35] Gerentea C, Andresa Y, McKayb G, Le Cloirecc P. Removal of arsenic(V) onto chitosan: from sorption mechanism explanation to dynamic water treatment process. Chem Eng J 2010;158:593–8.

[36] Pal P, Ahamad SkZ, Pattanayak A, Bhattacharya P. Removal of arsenic from drinking water by chemical precipitation- a modeling and simulation study of the physical-chemical processes. Water Environ Res 2007;79(4):357–66.

[37] Clifford DA. Ion exchange and in organic adsorption. In: Pontius FW, editor. Water quality and treatment. American Water Works Association. 4th ed. New York: McGraw Hill Inc; 1991.

[38] Fox KR, Sorg TJ. Controlling arsenic, fluoride and, uranium by point-of-use treatment. J Am Water Works Assoc 1998;81:94–101.

[39] Hathway SW, Rubel FJ. Removing arsenic from drinking water. J Am Water Works Assoc 1987;79:61–5.

[40] Chen HW, Feey MM, Clifford D, Mcneill LS, Edward M. Arsenic treatment considerations. J Am Water Works Assoc 1999;91:74–85.

[41] Lin TF, Wu JK. Adsorption of arsenite and arsenate within activated alumina grains: equilibrium and Kinetics. Water Res 2001;35(8):2049–57.

[42] Blanchin L. Ph.D. Thesis. University of Lyon; June 1952.

[43] Whittington B, Llievski D. Determination of the gibbsite dehydration reaction pathway at conditions relevant to Bayer Refineries. Chem Eng J 2004;98:89–97.

[44] Perry RH, Tien C. Adsorption calculations and modeling. Boston: Butterworth-Heinemann; 1994.

[45] McCabe WL, Smith JC, Harriott P. Unit operations of chemical engineering. 3rd ed. New York: McGraw-Hill; 1993.

[46] Moureau-Zabotto L, Thomas L, Bui BNG, Chevreau C, Stockle E, Martel P, et al. Management of soft tissue sarcomas (STS) in first isolated local recurrence: a retrospective study of 83 cases. Radiother Oncol 2004;73(3):313–9.

CHAPTER 4

Arsenic Removal by Membrane Filtration

Contents

Groundwater Arsenic Remediation
http://dx.doi.org/10.1016/B978-0-12-801281-9.00004-7

4.1 INTRODUCTION TO ARSENIC REMOVAL BY MEMBRANES

Separation and purification by membranes depend on two major principles, namely size exclusion and Donnan exclusion. In the former case, it is the sieving mechanism that separates target components from a solution or mixture based on relative sizes of the components of the solution or mixture and the pore size of the membrane. The Donnan exclusion principle, on the other hand, concerns the type and magnitude of the electrostatic charge on the surface of the membrane and the components of interest. Basically it is the charge repulsion phenomenon that causes the desired separation. Membranes are classified based on these principles and the driving force involved in effective separation. Table 4.1 thus helps the selection of a type of membrane based on the situation and requirement of separation. The target solute in the present case is arsenic in different forms as described in Chapter 1. Figures 1.1 and 1.2 show the presence and speciation of arsenic in groundwater. From these figures and Table 4.1, it appears that nanofiltration (NF) and reverse osmosis (RO) membranes with solution diffusion mechanism for solute transport have the highest potential of producing safe drinking water from the arsenic-contaminated groundwater.

Arsenic can be present in water in a particulate (>0.45 μm), colloidal (between 0.45 μm and 3000 Da), or dissolved state (<3000 Da). However, barely 10–20% arsenic is found to be in a colloidal or particulate state [1]. In groundwater, arsenic remains mostly in a dissolved state, thereby indicating that there is the possibility of very limited success of microfiltration

Table 4.1 Major Membrane Types and Membrane-Based Processes

Membrane process	Membrane type	Pore size (Å or nm)	Separation mechanism
Particle filtration	Porous	$>50,000$	Size exclusion
Microfiltration (MF)	Porous	500–50,000 Å (<50 nm)	Size exclusion
Ultrafiltration (UF)	Porous	20–50 Å (2–50 nm)	Size exclusion + Donnan exclusion
Nanofiltration (NF)	Partly porous	Average 1 nm	Size exclusion + Donnan exclusion
Reverse osmosis (RO)	Partly dense / Dense	<5 nm	Solution diffusion

(operating size range being 0.08–2.0 μm) and ultrafiltration membranes (operating size range being 0.005–0.02 μm), where the separation mechanism is based on the principles of size exclusion. Thus ultrafiltration alone can separate only 10% of arsenic [2], and even this insignificant separation is from charge repulsion due to the fact ultrafiltration membranes, like most nanofiltration membranes, bear a small negative charge. However, a combination of chitosan (an environment-friendly biopolymer) and humic acid (from naturally occurring humic substances or dissolved organic matter, DOM) can effectively remove 65% of arsenic from water [2,3]. Chitosan has weak affinity for arsenic but can strongly adsorb DOM. DOM with an average molecular weight of 35,000 Da or greater can cause chelation of arsenic and its subsequent removal by UF membranes. The technique however, cannot bring arsenic concentration levels down to 10 μg/L, and with a small charge on the surface, an ultrafiltration membrane can separate arsenic only to a very limited extent whereas microfiltration (pore size being of the order of 0.05 μm) can do so only with prior physicochemical treatment like coagulation and flocculation [4,5]. Thus from speciation and size distribution of arsenic in water and from the associated mechanisms of separation of the broad classes of membranes, use of ultrafiltration (UF) and microfiltration (MF) for removal of arsenic from contaminated groundwater is almost ruled out though under a hybrid treatment scheme, their use may be recommended, particularly for surface water.

By applying about 25 V of electric voltage on an ultrafiltration membrane, about 79% arsenic separation with reasonably high flux has been obtained [6]. However, the technique still falls short of target reduction below the WHO-prescribed limit of 10 μm/L. A study of Table 4.2 will further help in selecting the most appropriate type of membrane in arsenic separation from water.

Table 4.2 Driving Forces and Their Related Membrane Separation Processes

Driving force	Membrane process
Pressure difference	Microfiltration, ultrafiltration, nanofiltration, reverse osmosis or hyperfiltration
Chemical potential difference	Evaporation, per traction, dialysis, gas separation, vapor permeation, liquid membranes
Electrical potential difference	Electrodialysis, membrane electrophoresis, membrane electrolysis
Temperature difference	Membrane distillation (MD)

In RO and NF membranes, separation of target components and transport through membranes follow solution diffusion mechanisms where the components of interest either dissolve into the membranes or interact with membrane materials and migrate by molecular diffusion across the membrane to reemerge on the other side of the membrane.

Permeability of such dense membranes is related to solubility and diffusivity through the equation

$$P = S \times D \qquad (4.1)$$

where P = permeability of the membrane, S = solubility, D = diffusivity.

As the solute and solvent fluxes are uncoupled in the solution diffusion transport regime that dominates in RO and NF membranes, rejection increases with the increase of applied transmembrane pressure against a decreasing trend of rejection by MF and UF membranes as illustrated in Figure 4.1.

The operating pressure ranges for the major pressure-driven membranes are as indicated in Table 4.3. Because with increasing pressure solvent flux increases but rejection decreases in MF and UF membranes, a balance has to be struck in deciding on the operating pressure. During transport through NF and RO membranes, transmembrane pressure shows a strong positive

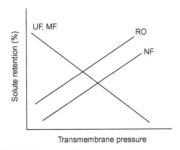

Figure 4.1 Rejection and flux behavior of major membrane types under varying pressure.

Table 4.3 Operating Pressure Across Membrane Types

Membrane types	Operating pressure (bar)
Microfiltration (MF)	<5
Ultrafiltration (UF)	5–10
Nanofiltration (NF)	10–20
Reverse osmosis (RO)	>20

correlation with both flux and rejection but considering mechanical strength of the membranes as well as membrane housing material, optimization has to be done with respect to the applicable pressure.

4.2 MEMBRANE FILTRATION MODES, MODULES, AND MATERIALS IN ARSENIC SEPARATION

4.2.1 Concentration polarization and fouling of membrane surface

Since fouling is considered one of the biggest hindrance factors in effective use of membranes in various separation and purification applications, this issue must be taken into account before selecting a type of membrane, membrane module, and mode of fluid flow since these are the three most significant factors that largely determine the extent of fouling of membrane surface. Fouling of the membrane surface occurs largely due to the phenomenon of adsorption that basically follows concentration polarization of solute molecules, which is the accumulation of solutes in a layer called the boundary layer near the membrane surface. Concentration of solutes in this boundary layer remains much larger than that in the bulk solution. In fact concentration polarization is the culmination of the equilibrium established between the fast process of convective diffusion of the solutes from the bulk solution to the membrane surface and the slow process of back diffusion of the solutes from the membrane surface to the bulk solution.

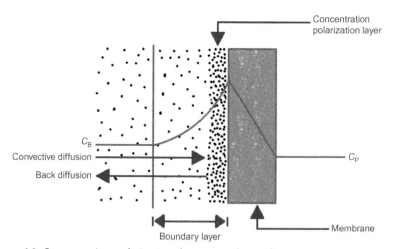

Figure 4.2 Concentration polarization during membrane filtration.

Figure 4.2 shows that due to concentration polarization, concentration of solutes at the surface (say, C_S) is much larger than concentration of the solutes in the bulk (C_B). This results in a percentage rejection as $R_{actual} = (1 - C_P/C_S) \times 100$ instead of theoretical rejection $R_{ideal} = (1 - C_P/C_B) \times 100$. Flux naturally decreases due to higher encountered resistance to fluid flow. Concentration polarization not only directly facilitates fouling of the membrane by the deposited solutes but also enhances chemical precipitation at the membrane surface due to increased solute concentration that often exceeds the solubility product of the solutes.

Figure 4.3 illustrates how deposited solutes may cause fouling of membranes through the clogging of membrane pores or channels. While the porous membranes are largely characterized by their pore size, pore distribution, and hydrophobicity, the nonporous or dense membranes are characterized by their electrostatic membrane charge density and permeability as expressed through Eq. (4.1). Membranes for arsenic separation have to be selected according to the arsenic speciation in the water from which it has to be removed.

Hydrodynamics, relative sizes of the solutes and membrane pores, electrostatic charges of the ions, and membrane surface are the governing parameters of concentration polarization, which in turn determines flux and rejection. Hydrodynamics can be made favorable to separation at a reasonable flux level. For example by using spacers in membrane modules, or by introducing cross-flow mode, turbulence can be enhanced, which in turn can reduce concentration polarization. It is obvious that greater drag force associated with greater flux causes increased concentration polarization as deposited solutes lose mobility under such high drag force. This is where the concept of critical flux assumes significance. This critical flux is the flux

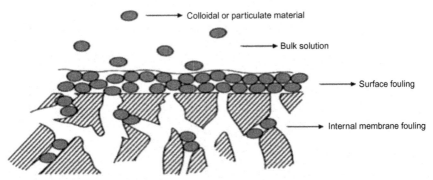

Figure 4.3 Typical fouling of membrane by deposited solutes.

below which no further decline in flux takes place with the progress of time of membrane filtration. This means that the higher the critical flux of a membrane filtration system, the better the performance of the system. A higher critical flux can be ensured by keeping the concentration polarization at a low level, which in turn is possible under higher cross-flow velocity (leading to better sweeping action) or higher charge repulsion.

4.2.2 Filtration Mode

A membrane device can be operated in batch mode, continuous mode, or feed-and-bleed mode. In batch mode, a given volume of feed is subjected to filtration under hydrostatic pressure when the solvent permeates primarily through the membrane, leaving behind the solute retained on the surface of the membrane. The process is terminated when the desired concentration of the solute on the feed side is reached. In a fully continuous mode of operation, the feed solution is continuously fed on the separating surface of the membrane and the desired filtrate is continuously collected from the other side of the membrane surface. It is assumed that the filtration process yields the desired concentrations on either the permeate side or the feed side, and the intended recovery is achieved through such continuous operation. However, if the desired recovery is not achieved in this once-through process, a part of the retentate or even the whole retentate-bearing solution is recycled back to the initial feed tank for multiple passes through the filtration surface of the membrane surface until the desired recovery is achieved. This is called the feed-and-bleed mode of filtration. Fluid may encounter the separating surface of the membrane in a direction perpendicular to the membrane surface as depicted in Figure 4.4. This is called the dead-end mode of filtration. Such a mode leads to rapid deposition of solid on the filtration surface. If the fluid encounters the membrane in a tangential fashion as depicted

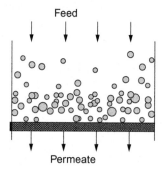

Figure 4.4 Dead-end mode of filtration.

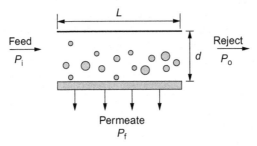

Figure 4.5 Cross-flow mode of filtration.

in Figure 4.5, the mode of operation is called cross-flow. In such a flow pattern, deposition of the solid on the membrane surface is strongly discouraged by virtue of the sweeping action of the fluid flow.

The flow of the bulk solution is parallel to the membrane surface. The solvent flow through the membrane carries particles to the surface, where they form a thin layer.

A relatively high flow rate tangential to the surface sweeps the deposited particles toward the filter exit leaving a relatively thin deposited cake layer similar to the gel layer formed in UF. This cross-flow is effective in controlling concentration polarization and cake buildup, allowing relatively high fluxes.

Transmembrane pressure in a cross-flow is expressed as

$$\Delta P_{TM} = \left(\frac{P_i - P_o}{2}\right) - P_f \qquad (4.2)$$

For *laminar flow*, this pressure drop following Poiseuille flow is

$$\Delta P = \frac{(C_1 \mu L v)}{d^2}$$
$$= \frac{(C_2 \mu L Q)}{d^4}$$

For *turbulent flow* (follow Fanning equation):

$$\Delta P = \frac{(C_3 f L v^2)}{d}$$
$$= \frac{(C_4 f L Q^2)}{d^5}$$

where C_1, C_2, C_3, $C_4 =$ constants based on channel geometry; $f =$ factor based on Reynolds' number; $Q =$ volumetric flow rate; $v =$ velocity; $L =$ filter length; $d =$ fluid channel height above the membrane; $\mu =$ viscosity.

In the dead-end flow model, the particles build up with time as a cake and the clarified permeate is forced through the membrane. Membrane resistance is constant, whereas cake resistance increases with time due to cake buildup.

The permeate flux equation under dead-end filtration mode may be written as

$$N_w = \frac{\Delta P}{\mu(R_m + R_c)} \qquad (4.3)$$

where N_w = solvent flux, kg/(m^2 sec); ΔP = pressure difference, Pa; R_m = membrane resistance, m^2/kg; R_c = cake resistance, m^2/kg; μ = viscosity of the solvent, Pa sec.

4.2.3 Membrane modules

The major membrane modules that have been used [7–12] in arsenic removal are plate and frame module, hollow-fiber module, spiral-wound module, shell and tube or tubular module, and flat-sheet cross-flow module.

4.2.3.1 Plate and frame module

In plate and frame module, membrane, feed spacers, and product spacers are layered together between two end plates. Arsenic-contaminated water is forced across the surface of the membrane. A portion passes through the membrane, enters the permeate channel, and makes its way to a central permeate collection manifold.

This module, as shown in Figure 4.6, permits easy cleaning and membrane replacement, and good flow controls on both permeate and feed side of the membrane. This module has been used in arsenic removal fitting nanofiltration membrane [8].

Such modules at the most are used for small-scale applications. These modules are expensive compared to the alternatives due to the large number of spacer plates and seals. The module, however, has a potential leakage problem—the gaskets required for each plate under fouling rises rapidly unless they are cleaned frequently, disrupting operation.

4.2.3.2 Flat-sheet cross-flow module

This relatively recent module has been used successfully not only in arsenic separation but also in many other important applications [7,13,14]. Flat-sheet cross-flow module is the best choice where fouling is the major concern. By virtue of the sweeping action of fluid over the membrane surface a long fouling-free operation is possible. The module (shown in Figure 4.7)

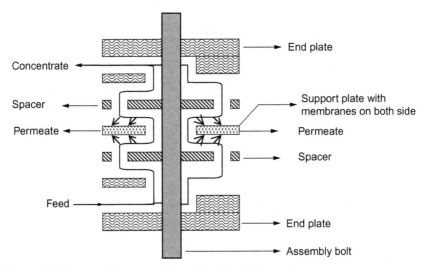

Figure 4.6 Plate and frame membrane module [12].

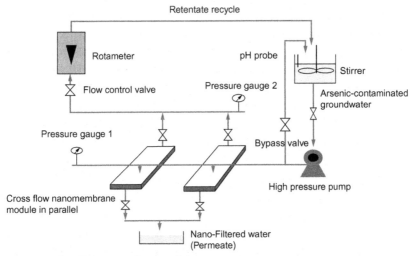

Figure 4.7 Flat-sheet cross-flow module.

yields high flux and offers ease of membrane replacement and cleaning like the plate and frame module.

4.2.3.3 Spiral-wound module

This design, as presented in Figure 4.8, consists of a membrane envelope of spacers and membrane wound around a perforated central collection tube. The module is placed inside a tubular pressure vessel. Feed passes axially down the module across the membrane envelope. A portion of the feed

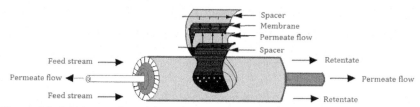

Figure 4.8 Spiral-wound module [12].

permeates into the membrane envelope, where it spirals toward the center and exits through the collection tube. The hydrodynamics can be controlled by changing the spacer thickness to overcome the concentration polarization and fouling. However, the membrane surface area per unit volume is low. The system may become expensive for high pressure applications because of the requirement of a high-pressure vessel for housing. Bypassing of feed may occur due to nonuniform wrapping of the spacer.

4.2.3.4 Hollow-fiber module

Hollow-fiber membrane modules (Figure 4.9) are formed in two basic geometries—one with a shell-side feed design and the other with a bore-side feed design. Shell-side feed modules are generally used for high-pressure applications well up to 1000 psig. In such a module, a loop or a closed bundle of fibers is contained in a pressure vessel. The system is pressurized from the shell side; permeate passes through the fiber wall and exits through the open fiber ends. Because the fiber wall must support considerable

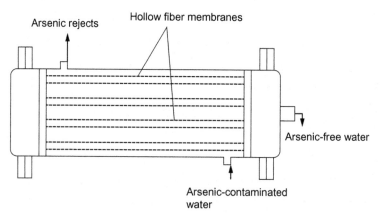

Figure 4.9 Hollow-fiber membrane module in arsenic removal [12].

hydrostatic pressure, the fibers usually have small diameters and thick walls, typically 50-μm internal diameter and 100- to 200-μm outer diameter. In the bore-side feed design, the fibers are open at both ends, and the feed fluid is circulated through the bore of the fibers. To minimize pressure drop inside the fibers, the diameters are usually larger than those of the fine fibers used in the shell-side feed system and are generally made by solution spinning. Feed pressures are usually limited to below 150 psig in this type of module.

The biggest advantage of the hollow-fiber module is the very large membrane surface area packed in a single module. The problem with the shell-side feed module is that fouling on the feed side of the membrane occurs frequently, necessitating pretreatment of the feed stream. Bore-side feed modules are generally used for medium-pressure feed streams up to 150 psig, for which good flow control to minimize fouling and concentration polarization on the feed side of the membrane is desired. The bore-side feed types are used in ultrafiltration, pervaporation, and some low-to-medium-pressure gas applications.

4.2.3.5 Shell and tube or tubular module

The tube diameter of this module as illustrated in Figure 4.10 is in the range of 5–15 nm. In a typical tubular membrane system a large number of tubes are manifold in series where the feed solution is pumped through all tubes connected in series. The permeate from each tube gets collected in a common collection header. The tubes consist of a porous paper or fiberglass support with the membrane formed on the inside of the tubes.

This module provides a large membrane area per unit volume but membrane formation is complex since the support and selecting layer are formed as an integral cylindrical unit during spinning. This module is fouling-prone. High volumetric holdup is another disadvantage. Tubular modules are now generally limited to UF applications, for which the benefit of resistance to membrane fouling due to good fluid hydrodynamics compensate their high

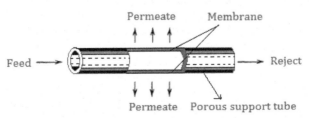

Figure 4.10 Shell and tube or tubular module.

cost; that is, this module is used when turbulent flow regime is preferred due to high solid concentration of the fluid ($Re > 10,000$). Thus use in purification of arsenic-contaminated groundwater is not recommended.

4.2.4 Membrane materials

While UF membranes used in removal of arsenic in very limited cases are basically sulfonated polysulfone and hydrophobic type having small negative charge on the surface, the nanofiltration and reverse osmosis membranes used are aromatic polyamide, composite polyamide, polyether urea, thin film composite, polyvinyl alcohol, or cellulose acetate types manufactured by Desal, Sepro Membranes, Film Tec, Hydraunautics, Fluid systems, GE Osmonics, Nitto Electric Industrial Company, and others. The composite nanofiltration membranes may be having asymmetric pore structure with a thin skin layer as depicted in Figure 4.11. A porous structure provides the mechanical strength.

An RO membrane with asymmetric pore structure is normally having a dense skin layer as illustrated in Figure 4.12. The top skin layer in both cases act as the main screening layer.

Major characteristics of a set of typical nanofiltration membranes used in arsenic separation are presented in Table 4.4.

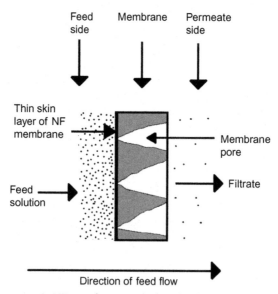

Figure 4.11 Asymmetric NF membrane with thin skin layer.

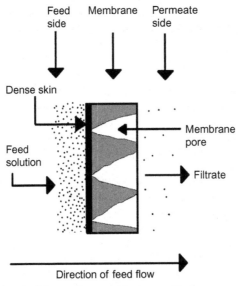

Figure 4.12 Asymmetric RO membrane with dense skin layer.

Table 4.4 Properties of the Composite, Flat-Sheet Polyamide Nanofiltration Membranes Used in Arsenic Removal at 14–16 Bar Pressure
Major characteristics and performance indicators of the investigated membranes at 12–16 bar

Characteristics	NF-1	NF-2	NF-3
Module type	Cross-flow	Cross-flow	Cross-flow
Membrane geometry	Flat-sheet	Flat-sheet	Flat-sheet
Membrane material	Polyamide	Polyamide	Polyamide
Thickness (cm)	0.0165	0.0165	0.0165
Temp (°C) resistance	50	50	50
pH resistance	2–11	2–11	2–11
Pure water flux $(L/(hr\,m^2))$	110	130	120
Acetic acid rejection (%)	84	12	35
Arsenic (As(V))	98–99	94–95	95–96
Investigated solute rejection (%)			
$MgSO_4$	99.5	97	98
NaCl	95	85	83
$MnSO_4$	97	92	95

Sources: References [7,13,14].

4.3 LOW-PRESSURE MEMBRANE FILTRATION IN ARSENIC REMOVAL

MF and UF membrane modules are operated under low pressure. Though use of MF and UF membranes is not recommended in general for removal of arsenic in view of limited possibility of separation, they may sometimes be used for surface water containing arsenic in particulate form. Because the pore size of MF membranes is large, it can remove particulate forms of arsenic from water. Therefore, the arsenic removal efficiency by MF membrane is highly dependent on the size distribution of arsenic-bearing particles in the water. In order to increase the arsenic removal efficiency by MF, coagulation and flocculation processes, which can increase arsenic particle size, are adopted to assist the MF techniques.

Coagulation and MF combined (FMF) is one such technique for arsenic removal. The coagulation process consists of the addition of an iron-based coagulant, such as ferric chloride when $FeCl_3$ hydrolyzes in water to form $Fe(OH)_3$ precipitate. $Fe(OH)_3$ precipitate carries a net positive charge on the surface. In the pH range of 4.0–10.0, negative charge-carrying arsenate ions adsorb onto the positively charged $Fe(OH)_3$ and get separated following the mechanism discussed in Chapter 2. Over this pH range, arsenite remains in neutral molecular form. For effective removal of arsenic from water a complete oxidation of arsenite to arsenate is required as the hydroxide coagulation process depends mainly on ionic interactions. The combination of flocculation and MF may culminate in a new technique called FMF. The arsenic removal efficiency using FMF is higher than removal using MF. This FMF process has been found [15] to produce treated water containing less than 2 µg arsenic/L from feed water containing 40 µg arsenic/L. Coagulation combined with microfiltration can be very effective in removing up to 99% of arsenic from surface water [16].

In case of UF membranes, success, though limited, depends on whether the UF membrane is charged or uncharged. The other important factors that determine separation efficiency are the quantity of DOM, pH value, and speciation of arsenic. The negatively charged UF membrane indeed is more effective for As(V) removal than the uncharged UF membrane. The higher removal rates are results of electrostatic interaction between arsenic ions and negatively charged membrane surface. The dominant mechanism of separation by UF remains the sieving mechanism.

4.4 HIGH-PRESSURE MEMBRANE FILTRATION IN ARSENIC REMOVAL

There are two broad types of high-pressure membrane processes described to be effective in removing arsenic from water, namely nanofiltration (NF) and reverse osmosis (RO). RO is one of the oldest membrane technologies and also has been identified as a likely best available technology applied for small water treatment systems to remove arsenic from water.

In removing arsenic from water, both RO and NF have been most widely used. Very effective removal of arsenic, up to 95%, has been demonstrated [1,17,18] using RO and NF membranes. For the same rate of arsenic removal, NF can yield much higher flux than RO. In other words, it can be said that for the same flux and rejection, a NF can be operated at much lower pressure.

4.4.1 Reverse osmosis in arsenic removal

The chemical potential of a pure solvent is always higher than the chemical potential of the solvent in solution. So if a RO membrane, being a semipermeable membrane (permeable to solvent but impermeable to solute), separates a pure solvent and a solution of the same solvent, then in osmosis, a spontaneous transport of solvent occurs from the pure solvent side to the solution side; in other words, from the dilute solution side to the concentrated solution side across the semipermeable membrane.

The chemical potential of a solvent in solution may be expressed in terms of activity of the liquid solvent (a) and chemical potential of the solvent in a pure state as

$$\mu_s^{solution} = \mu_s^0 + Vp + RT\ln a \tag{4.4}$$

where $\mu_s^{solution}$ is the chemical potential of the solvent in solution, μ_s^0 is the chemical potential of the solvent in a pure state, R is a universal gas constant, and T is the absolute temperature of the solution. V is partial molar volume and p is pressure. The solvent molecules can pass through the semipermeable membrane in both directions but the net flow is obviously from dilute to concentrated solution side under normal osmosis, as long as the levels of the liquids on both sides are the same, implying that the hydrostatic pressure on both sides is the same. The solvent flow from dilute to concentrated solution side can be reduced by exerting a pressure on the salt-solution side of the membrane. The pressure required to stop the solvent flow in this direction is called the osmotic pressure (π) of the salt solution, and an equilibrium is reached at this stage when the amount of solvent passing in opposite directions is just equal. The chemical potentials of the solvent on both sides of the membrane are equal. To reverse the flow of the water so that it flows from the salt solution to the fresh solvent, the pressure is increased above the osmotic pressure on the solution side as illustrated in Figure 4.13. This phenomenon is called reverse osmosis.

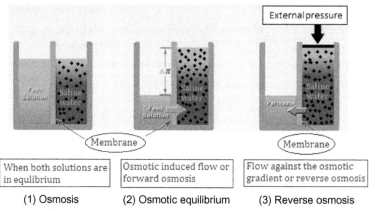

When both solutions are in equlibrium	Osmotic induced flow or forward osmosis	Flow against the osmotic gradient or reverse osmosis
(1) Osmosis	(2) Osmotic equilibrium	(3) Reverse osmosis

Figure 4.13 Normal and reverse osmosis phenomena.

RO can operate at ambient temperature without phase change. Osmotic pressure is negligible in UF, NF, and MF. The osmotic pressure π of a solution is proportional to the concentration of the solute and temperature T. For dilute aqueous solutions,

$$\pi = \frac{n}{V_m} RT \tag{4.5}$$

where $n =$ the number of kmol of solute, $V_m =$ the volume of pure solvent water in m^3 associated with n kmol of solute, $R =$ the gas law constant $= 82.057 \times 10^{-3}$ (m^3atm./(kmol K)), and $T =$ temperature, K.

If a solute exists as two or more ions in solution, n represents the total number of ions. Osmotic pressure (psi) is

$$\pi = 1.12(T + 273)\sum m_i \tag{4.6}$$

where $T =$ temperature, °C, and $\sum m_i =$ summation of molalities of all ionic and nonionic constituents in solution.

4.4.1.1 Mass transport through RO membrane

There are two basic mass-transport mechanisms, diffusion mechanism and sieving mechanism. In diffusion mechanism, both the solute and the solvent migrate by molecular diffusion in the polymer, driven by concentration gradients set up in the membrane by the applied pressure difference. In sieve-type mechanism, the solvent moves through the micropores in essentially viscous flow, and the solute molecules small enough to pass through the pores are carried by convection with the solvent. Diffusion of the solvent through the membrane is illustrated in Figure 4.14.

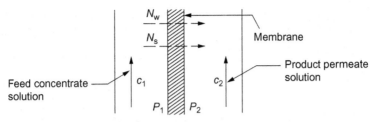

Figure 4.14 Concentration and fluxes in RO.

$$N_w = \frac{P_w}{L_m}(\Delta P - \Delta \pi) = A_w(\Delta P - \Delta \pi)$$

$$A_w = \frac{P_w}{L_m}$$

(4.7)

where N_w = the solvent (water) flux, kg/(m^2 sec); P_w = the solvent membrane permeability, kg solvent/(m atm. sec) L_m = the membrane thickness, m; A_w = the solvent permeability constant, kg solvent/(m^2 atm. sec); ΔP = hydrostatic pressure difference, atm.; P_1 = pressure on feed side, atm.; P_2 = pressure on product side, atm. $\Delta \pi$ = osmotic pressure of feed solution − osmotic pressure of product solution, atm.; subscript 1 ⇒ feed or upstream side of the membrane; subscript 2 ⇒ product or downstream side of the membrane.

Solute flux for the diffusion of the solute through the membrane may be expressed as

$$N_s = \frac{D_s K_s}{L_m}(c_1 - c_2) = A_s(c_1 - c_2)$$

$$A_s = \frac{D_s K_s}{L_m}$$

(4.8)

where N_s is the solute flux, kg solute/(m^2 sec); D_s is the diffusivity of solute in membrane, m^2/sec.

$$K_s = \text{the distribution coefficient} = \frac{c_m}{c} = \frac{\text{concentration } of \text{ solute in membrane}}{\text{concentration of solute in solution}}$$

(4.9)

A_s = the solute permeability constant, m/sec; c_1 = the solute concentration in upstream or feed (concentrate) solution, kg solute/m^3; c_2 = the solute concentration in downstream or product (permeate) solution, kg solute/m^3; the distribution coefficient K_s is approximately constant over the membrane.

4.4.1.2 Steady-state material balance for solute

The solute diffusing through the membrane = the amount of solute leaving in the downstream or product (permeate) solution:

$$N_s = \frac{N_w c_2}{c_{w2}} \tag{4.10}$$

where c_{w2} = the concentration of solvent in stream 2 (permeate), kg solvent/m^3. If the stream 2 is dilute in solute, c_{w2} is approximately the density of the solvent. Solute rejection may be expressed as

$$R = \frac{c_1 - c_2}{c_1} = 1 - \frac{c_2}{c_1} \tag{4.11}$$

4.4.2 Arsenic removal by nanofiltration

Water with very high arsenic content can also be treated by nanofiltration membrane without necessitating high transmembrane pressure. When a solution containing ions like arsenate is brought in contact with a membrane possessing a fixed like surface charge, the passage of ions through the membrane is inhibited due to the Donnan effect [19]. So, the rejection of arsenate will be higher if the selected membranes are negatively charged, and the rejection of anions will be higher compared to neutral arsenite. During separation of solutes by nanofiltration membranes, both the sieving mechanism (size exclusion) and the Donnan mechanism (electrostatic charge repulsion) may play their role. However, at higher pH (greater than 7.0) most of the nanofiltration membranes of polyamide composite types possess negative zeta potential [20]. When membrane surface potential measured as zeta potential changes from positive to negative ones with high medium pH, electrostatic interaction, charge repulsion, and hence separation of ionic species based on the Donnan effect become dominant in nanofiltration. Arsenic may occur in various oxidation states (−3, 0, +3, +5) and speciation may change from neutral to anionic forms and vice versa depending on the pH of the medium. Thus polyamide-type NF membranes, which largely assume negative surface potential (at pH value greater than 7.0), can be effectively exploited in removal of arsenic from water provided arsenic speciation is favorably changed, making the Donnan mechanism the dominant separation mechanism. However, efficient removal of arsenic by NF depends on a host of factors like types of membranes, modules of membranes, operating pressure, pH, oxidation states of arsenic, concentration of arsenic and, presence of other impurities like iron and such in water.

A very few studies have been reported on nanomembrane separation of arsenic using some specific membrane modules and simulated arsenic solution. Varieties of other modules with different types of membranes have the potential of arsenic separation from groundwater offering longer fouling-free operation of the polymeric membranes and thus adding economy to the process. Such modules need to be tested using actually contaminated groundwater for judging the potential of the NF process in effective separation of arsenic in the presence of other possible contaminants. In the present study, investigations were carried out using a few commercially available polyamide composite NF membranes in a cross-flow module while using contaminated groundwater from some affected areas of the Bengal Delta basin in India.

Nanofiltration membranes as depicted in Figure 4.11 have asymmetric structure with a microporous or nearly dense skin layer over a porous support structure. Mass transfer through nanofiltration is the culmination of diffusion through a membrane matrix accompanied by viscous flow through the channels or pores. The governing factors in transport through NF will be described in the modeling section. Composite polyamide nanofiltration membranes are found to effectively separate arsenic from groundwater whenever conditions conducive for Donnan exclusion prevail. Typical composite nanofiltration membranes useful in this application may be well represented by the NF-1 membranes manufactured by Sepro Membranes Inc., USA as presented in Table 4.4. A little examination of this table shows that NF-1 having high negative membrane charge density is likely to be the most effective in arsenic removal exploiting the Donnan exclusion principle. NF membranes are found to be very effective [14] in the removal of other contaminants that may be present in arsenic-contaminated groundwater as indicated in Table 4.5.

Table 4.5 shows typical arsenic-contaminated water samples collected from the arsenic-affected Bengal Delta basin of India, which has the largest concentration of arsenic-affected population in a single geographic location of the world. Water samples in Table 4.5 show significant presence of iron, which often stands as a major hurdle in adsorption-based water purification. However, during nanofiltration, presence of other contaminants may not always have an antagonistic effect in rejection of arsenic. We shall discuss these aspects in connection with hybrid treatment later. Selection of an appropriate module is very important in ensuring long-term operation of the filtration device without significant reduction in flux due to deposition of other rejected contaminants. One such successful NF membrane module is a flat-sheet cross-flow module in feed-and-bleed operational mode as illustrated in Figure 4.15.

Table 4.5 Characteristics of Groundwater Samples from Three Different Locations of West Bengal, India

Parameters	Sample without arsenic	Sample I Chakdah, Nadia	Sample II Debnagar, Murshidabad	Sample III Sangrampur, N-24 Parganas
Total hardness (mg/L) (As $CaCO_3$)	403	538	400	650
Total alkalinity (mg/L) (As $CaCO_3$)	367	390	320	520
Chloride (mg/L) (As Cl^-)	320	310	270	310
Sulfate (mg/L) (As SO_4^{2-})	3	3	5	3
pH	7.10	7.2	7,6	7.15
Total Fe (mg/L)	2.5	4.2	10.5	7.5
Total As (μg/L)	Nil	150	376	252
As(III)	Nil	62	152	98
As(V)	Nil	88	224	154

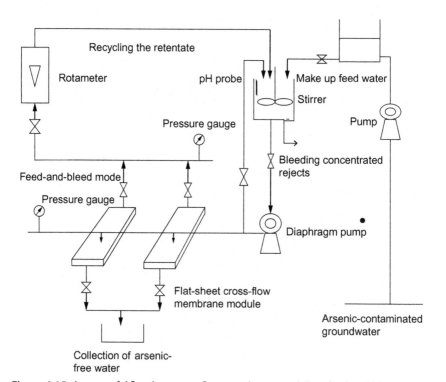

Figure 4.15 A successful flat-sheet cross-flow membrane module in feed-and-bleed mode.

4.4.3 Filtration process in flat-sheet cross-flow module

A number of flat-sheet cross-flow membrane modules may be connected to the high pressure feeding pump that pumps and circulates water at around 12 to 16 bar pressure through the membrane modules. Since one pass is not sufficient in desired removal of the contaminants, multiple-pass or feed-and-bleed mode is normally adopted. Tangential fluid flow over the separating surface of the membrane modules results in a great sweeping action, which prevents the accumulation of the retained solids on the surface. This ensures almost fouling-free operation for an extended period. After several days' operation, simple water rinsing of the membranes can restore the original flux to a great extent. A diaphragm type of pump may be very suitable for such a system that offers quite a high discharge pressure, although the operating pressure requirement here is much less than what is required in reverse osmosis. Membranes can be easily fitted to the module by unscrewing the nuts and bolts and fixing the sheet membrane on the perforated sieve support placed in the grooves of the stainless steel or polycarbonate module box. Pulsating flow motion is an additional advantage in preventing fouling. The system can be operated continuously for a long time without the necessity of frequent membrane replacement or cleaning. When the arsenic rejects attain a high concentration level in the circulating tank, the same may be bled out of the system. Typically a polyamide composite NF membrane such as NF-1 of Sepro Membranes Inc., USA, can remove almost 98–99% As(V) with a high flux of 140–150 LMH (liter per square meter per hour) at 16 bar pressure. Where contaminated water contains both trivalent and pentavalent forms of arsenic, total rejection of arsenic may not exceed 65% as trivalent arsenic remains largely in neutral molecular form and escapes the dominating separation mechanism that is the Donnan exclusion.

4.4.4 System performance measurement

Three major performance indicators that are measured routinely are rejection, flux, and fouling. Rejection is measured using Eq. (4.11). To use this equation, concentrations of arsenic in feed water (C_1) and in the filtrate (C_2) are estimated first using any standard spectrophotometric method. An atomic absorption spectrophotometer is one such instrument capable of measuring arsenic in water samples down to the ppb level. In the spectrophotometric method, measurement is done at wavelength 193.7 nm. Using the flame-FIAS technique [21], arsenic is measured after its conversion to volatile hydride. Arsenic in sample water is prereduced (As(V) to As(III)) using a reducing solution 5% (w/v) potassium iodide (KI) and 5% (w/v) ascorbic acid. In this flame-FIAS technique, oxy-acetylene flame is used to atomize the sample element and FIAS (flow injection with atomic spectroscopy) is used to inject an exactly reproducible volume of the sample into a continuously flowing carrier system. After measurement of total arsenic, measurement of As(V) is kept suppressed by adjusting the pH of the mixed solution to 3 using sodium hydroxide and a citric acid buffer enabling measurement of As(III) only [22]. The measured As(III) is then deducted from total arsenic, giving the measurement for As(V). Percentage removal of arsenic is then calculated using Eq. (4.11).

Flux can be measured over time directly from the collected filtrate volume and the same is expressed in standard LMH units, representing liter per square meter membrane surface area per hour. A flux value of 50 LMH may be considered reasonably acceptable for operating a module for large-scale filtration. Requirement of membrane cleaning or replacement of membrane can be assessed from measuring and continuous monitoring of the flux. Rapid decline in flux indicates high degree of membrane fouling whereas insignificant flux decline over time in terms of days or even months indicates high potential of the membrane and the module in successful use for the filtration. In nanofiltration of arsenic-contaminated groundwater, flat-sheet cross-flow membrane modules are found to do the filtration work without any significant flux decline over weeks of operation. Presence of other contaminants like iron sometimes causes flux decline. Even pH is found to play significant role in flux and rejection. These aspects will be discussed in the next section.

Membrane morphology and membrane fouling can also be studied through SEM analysis (scanning electron microscopy). The membranes are freeze-fractured in liquid nitrogen. After a thin gold coating in ion sputter (Hitachi, E-1010) the pieces of membranes are transferred to the scanning electron microscope. SEM analysis at 20 kV may be done for the top surfaces of the membranes. Observation and comparison of the surface morphology before and after application can indicate extent and nature of fouling (Figures 4.16 and 4.17, respectively).

4.4.5 Flux behavior under varying transmembrane pressure

Figure 4.18 illustrates variation of flux with operating pressure. Water flux is found to vary linearly with transmembrane pressure irrespective of whether groundwater contains arsenic or not. This is so because dissolved arsenic leaves hardly any deposit on the separating membrane surface. Such behavior is the only indicator of nanofiltration suitability in treating

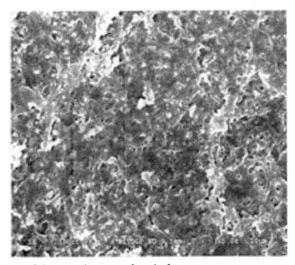

Figure 4.16 SEM of the membrane surface before use.

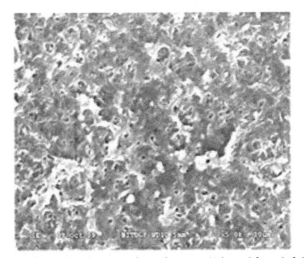

Figure 4.17 SEM of the membrane surface after use. *(Adapted from Ref. [7].)*

arsenic-contaminated groundwater. Flux varies almost linearly with pressure; the tightest type of the NF membrane shows the minimum flux.

4.4.6 Transmembrane pressure and rejection of arsenic

Figure 4.19 illustrates that transmembrane pressure has positive correlation with arsenic rejection. Arsenic rejection increases slightly with increase of

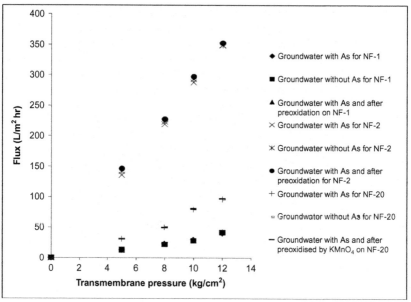

Figure 4.18 Relation between flux and transmembrane pressure. Symbols: ◆ Groundwater with As for NF-1; ▪ Groundwater without As for NF-1; ▲Groundwater with As and after preoxidation on NF-1; × Groundwater with As for NF-2; * Groundwater without As for NF-2; • Groundwater with As and after preoxidation for NF-1; + Groundwater with As for NF-20, – Groundwater without As for NF-20; — Groundwater with As and preoxidised by $KMnO_4$ on NF-20. Operating conditions: pH 7, pressure range 5–12 kgf/cm², temperature 35 °C, flow rate 700 L/hr, cross-flow velocity 1.16 m/sec. *(Adapted from Ref. [7].)*

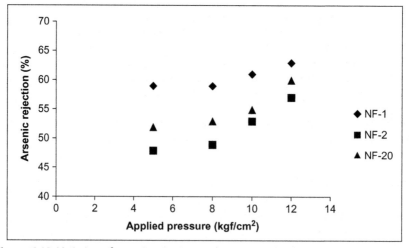

Figure 4.19 Variation of arsenic rejection with transmembrane pressure [7].

applied pressure. This may be attributed to the solution diffusion mechanism that applies to nanofiltration. In solution diffusion mechanism, solute flux and the solvent flux are uncoupled as a result, the increase of solvent flux following an increase in transmembrane pressure does not result in increase of solute flux. Rather, an increase of solvent flux stands in the way of the transport of solute. Transmembrane pressure leads to the increase in solvent flux. Presence of As(V) as monovalent or divalent anionic forms in water results in a charge repulsion when it comes in contact with negatively charged NF membranes, and this results in rejection of As(V). Due to the neutral character of the As(III) molecule within the pH range of 3–10, increase of pressure increases the flux. This in turn, decreases the As(III) retention since in this case, it is the convective transport that becomes dominant due to the neutral character of As(III) that passes through the membrane following size exclusion principle.

4.4.7 Role of medium pH during nanofiltration of arsenic-contaminated water

Figure 4.20 shows that with increase of pH of groundwater (with 150 μg/L arsenic) from 3.0 to 10.0, arsenic rejections increased from 50% to 76% for NF-1, 33% to 69% for NF-2, and 43% to 71% for NF-20 membranes, respectively, without preoxidation. But increase of pH along with preoxidation of arsenic species resulted in much higher arsenic rejection.

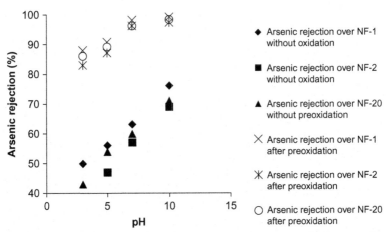

Figure 4.20 Effect of pH on removal of arsenic over NF-1, NF-2, and NF-20 membranes with and without preoxidation. Operating conditions: arsenic concentration 150 μg/L, transmembrane pressure 12 kgf/cm^2, flow rate 700 L/hr, cross-flow velocity 1.16 m/sec, temperature 35 °C, preoxidation done by KMnO$_4$ [7].

Rejections of arsenic at pH 7 (normal groundwater pH), for NF-1, NF-2, and NF-20 membranes were 57%, 60%, and 63%, respectively, when operated at a pressure of 12 kgf/cm^2. Arsenic rejection was 96–98% for all the three types of membranes at pH 7.0. At pH 10.0, arsenic rejection increased up to 99% for preoxidized water. pH has a significant role in removal for arsenic since speciation of arsenic changes with the pH of the medium. Up to a pH value of 8.0, As(III) remains largely as a neutral molecule while As(V) remains as an anion. As(V) speciation even changes from monovalent ($H_2AsO_4^-$) to divalent form ($HAsO_4^{2-}$), enhancing further retention of arsenic due to the Donnan exclusion. As the NF membranes are mostly negatively charged, rejection of arsenic seems to be affected by the charge valence of arsenate in the solution (Donnan exclusion). Apparent pore size of polyamide NF membranes can also vary with solution pH. At the pore surface point of zero charge (isoelectric point), the membrane functional groups are minimal in charge and hence open up, as the absence of repulsion forces contribute to the widening of the membrane pores. At high or low pH values, functional groups of membrane polymer can dissociate and take on positive or negative charge functions. Repulsion between these functions in the membrane polymer reduces or closes up membrane pores. Braghetta [23] has shown the effect of solution pH and ionic state on apparent pore size of membranes. At high ionic strength and high pH, apparent pore size reduces remarkably.

4.4.8 Iron in arsenic-contaminated water and its role in arsenic removal during nanofiltration

Feed water for the present investigation was collected from three different arsenic-affected areas of West Bengal in India. Characteristics of such feed waters in terms of impurities like salts, iron, and arsenic were found to vary from source to source as shown in Table 4.5. As(III) in feed water was oxidized by addition of $KMnO_4$ following the procedure of Pal et al. [21]. Nanofiltration of the preoxidized water was then performed using the NF-1 membrane, which was already found to be the most effective in arsenic removal. Figure 4.21 shows that with the increase in iron concentration in the groundwater, the arsenic separation efficiency of the membrane increased. A similar positive effect of Nano scale zero valent iron on removal efficiency during microfiltration and nanofiltration of arsenic-contaminated water is observed [10].

The presence of iron in arsenic-contaminated water as observed in the water samples of Table 4.5 poses a real problem in treatment of

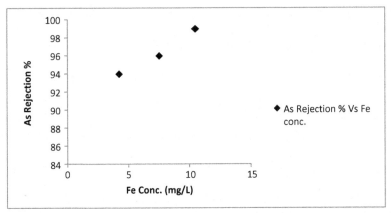

Figure 4.21 Effect of iron concentration in groundwater on arsenic removal efficiency of NF-1. Operating conditions: transmembrane pressure 12 kgf/cm^2, flow rate 700 L/hr, cross-flow vel 1.16 m/sec, contaminated water as per Table 4.1, temperature 35 °C, preoxidation done by KMnO$_4$ [7].

contaminated water by adsorption. Higher rejection of arsenic during nano-filtration in the presence of iron as an impurity may be attributed to enmeshed coprecipitation of arsenic with ferric hydroxide. Pentavalent arsenic precipitates out from water through adsorption, enmeshment, and formation of loose As(V)–Fe(OH)$_3$ complexes [21]. At the same time it is found that flux decreases with an increase in iron concentration as exhibited by Figure 4.22. Formation of such precipitates lead to flux reduction of water.

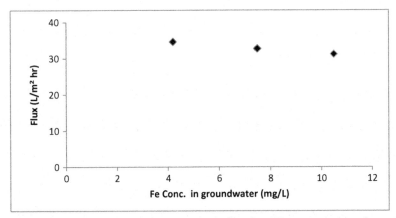

Figure 4.22 Effect of iron concentration on flux by NF-1 membrane. Operating conditions: transmembrane pressure 12 kgf/cm^2, flow rate 700 L/hr, cross-flow velocity 1.16 m/sec, temperature 35 °C, preoxidation done by KMnO$_4$ [7].

About 10% reductions in flux is observed as iron concentration increases from 4.2 to 10.5 mg/L. Such flux reduction can be attributed to possible clogging of the membrane by the precipitates.

RO membrane with a dense pore structure (pore size of about 0.0005 μm) has the capability of production of largely arsenic-free water, but high pressures are required to cause water to pass across the membrane from a concentrated to dilute solution. In general, driving pressure increases as selectivity increases. Both in RO and NF, it is desirable to achieve the required degree of separation at the maximum specific flux (membrane flux/driving pressure). Nanofiltration is based on the use of membranes constructed of a porous inert layer of polysulfone and a negatively charged hydrophobic rejection layer.

The transport of the solvent is accomplished through the free volume between the segments of the polymer of which the membrane is constituted, osmotic pressure becomes greater in the RO system compared to other processes, and the rate at which water diffuses across the membrane is very low.

4.5 HYBRID PROCESSES

When arsenic is present in water (particularly surface water) in particulate form, MF and UF can only partially remove arsenic from water. NF and RO are effective in removing arsenic in both particulate as well as dissolved states. RO membranes demand much higher operating pressure and hence involve both higher capital as well as operating costs than the NF membranes, though in removal efficiency RO is slightly better than NF membranes. However, NF is found to be most successful when arsenic is present in pentavalent form. Thus it transpires that NF membranes could possibly be the best option in arsenic removal from contaminated groundwater provided all trivalent arsenic is converted into pentavalent form. This indicates that no single purification treatment, chemical- or membrane-based, can alone purify arsenic-contaminated water down to the WHO-prescribed limit of 10 ppb level. Membrane-based hybrid processes have thus stepped in.

Such hybrid processes are of the following broad types:
- Adsorption integrated with ultrafiltration
- Coagulation–oxidation–membrane filtration
- Chemical oxidation integrated with flat-sheet cross-flow nanofiltration
- Chemical oxidation–nanofiltration integrated with chemical stabilization
- Nanofiltration/RO combined with chemical treatment

- Micro and ultrafiltration with zerovalent iron
- Electro-ultrafiltration

4.5.1 Adsorption integrated with ultrafiltration

Adsorption combined with ultrafiltration has been found to remove both chromium and arsenic from water. Arsenic from an initial concentration of 1 ppm can be brought down to 10 µg/L using Fe_2O_3 adsorbent nanoparticles [24]. In this process, adsorbent nanoparticles are totally removed in subsequent ceramic ultrafiltration membranes. However, in this process, removal of trivalent arsenic is not effective and frequent membrane fouling necessitates frequent regeneration of the ceramic membranes.

4.5.2 Coagulation–oxidation–membrane filtration

With prior coagulation of arsenic-bearing surface water with ferric chloride ($FeCl_3$) and ferrous sulfate ($Fe_2(SO_4)_3$), arsenic can be removed quite effectively. With prior oxidation by potassium permanganate ($KMnO_4$) and coagulation with ferrous and ferric sulfates arsenic concentration could be brought down from 200 to 300 µg/L to below 10 µg/L using a ZW-1000 hollow fiber membrane module [25].

4.5.3 Chemical oxidation integrated with flat-sheet cross-flow nanofiltration

Arsenic present in both trivalent and pentavalent form can be very efficiently removed in this largely fouling-free flat-sheet cross-flow membrane module [13]. A continuous stirred tank chemical reactor ensures chemical oxidation of trivalent arsenic into pentavalent form, facilitating almost total removal of arsenic from water in the downstream flat-sheet cross-flow nanofiltration module.

4.5.4 Chemical oxidation–nanofiltration integrated with chemical stabilization

This is a complete scheme of arsenic removal from water and its subsequent stabilization, which solves the disposal problem [14]. In the first stage, all trivalent arsenic is converted into pentavalent arsenic using $KMnO_4$ as an oxidizing agent. The second step comprises a flat-sheet cross-flow nanofiltration membrane module that removes over 98% arsenic in a continuous scheme without inviting the problem of membrane surface fouling over long periods of operation.

4.5.5 Nanofiltration/RO combined with chemical treatment

A hybrid treatment scheme may also consist of nanofiltration/reverse osmosis in the first stage, followed by chemical treatment with $Ca(OH)_2$ and H_2S with a volume reduction approach.

4.5.6 Micro- and ultrafiltration with zerovalent iron or chemical precipitation

It has been observed [10] that a dramatic increase in arsenic separation efficiency of micro- and nanofiltration membranes (Nitto Denko Corp., Japan) is possible with the addition of a small amount of zerovalent iron to feed water that contains a high level of arsenic (500 µg/L arsenic). Marked differences are noted, however, in separation efficiency with respect to trivalent and pentavalent arsenic species. Chemical precipitation of arsenic in water followed by microfiltration and ultrafiltration for separation of the precipitates is another hybrid process, the success of which depends largely on the chemical precipitation step rather than on membrane filtration.

4.5.7 Electro-ultrafiltration

Negatively charged arsenic species, As(V), is readily removed after applying voltage to the electro–ultrafiltration system (EUF cell). As(III) is removed via EUF after the pH of the water has been adjusted. Rejection of humic substances from water also increases due to the presence of an electric field. The removal of arsenite (III) from water relies primarily on electrostatic and nonelectrostatic mechanisms. In the presence of HSs, arsenate (V) complexes with the HSs and is then removed by electro-ultrafiltration [26].

4.6 CHEMICAL OXIDATION INTEGRATED WITH FLAT-SHEET CROSS-FLOW NANOFILTRATION

4.6.1 Oxidation–nanofiltration principle

Separation mechanisms in nanofiltration involve both steric (sieving) effects and electrical (Donnan) effects. This combination allows NF membranes to be effective for removal of more than 98% of arsenic from contaminated groundwater [7]. Lower pumping cost and membrane cost compared to RO makes nanofiltration an economically attractive option. Separation of ionic species by a nanofiltration membrane strongly depends on the membrane properties (membrane charge and membrane pore radius). A membrane with smaller pores is better able to retain ionic species, where a highly charged membrane is better able to exclude co-ions (ions of like charge as the membrane). When charge repulsion dominates separation, it is desirable that all the target species are converted into charged forms. As discussed in Chapter 1, arsenic may be present in groundwater both in trivalent and pentavalent forms. Under normal pH conditions, trivalent arsenic largely remains in neutral form, which cannot be separated by a Donnan exclusion using a nanofiltration membrane. However, conversion of trivalent arsenic into pentavalent form facilitates its separation by nanofiltration, as corresponding ions of arsenic being negatively charged get rejected by the membrane because of charge repulsion. Soluble arsenic can be effectively removed from groundwater by a NF membrane with prior oxidation of trivalent arsenic to pentavalent state using some oxidizing agent.

4.6.2 Performance and limitation of arsenic oxidants

Oxidants, such as ozone, hydrogen peroxide, chlorine, and potassium permanganate are the possible oxidants for conversion of As(III) to As(V). Certain merits and demerits are associated with each of these oxidants.

4.6.2.1 Ozone (O_3) as oxidant

For simultaneous oxidation of As(III) to As(V) and disinfection of water, ozone may be the most ideal choice. The major advantage in using ozone is that it leaves no harmful by-products. When ozone is added to water containing both trivalent arsenic as well as soluble iron, it oxidizes both forming sites on the ferric hydroxide for subsequent adsorption of arsenic. The arsenic-bearing iron hydroxide can then be removed by solid liquid separation processes. Ozone preoxidation before nanofiltration could present a problem if the organic carbon that is formed has a low molecular weight and passes through the membrane.

4.6.2.2 Hydrogen peroxide (H_2O_2)

Hydrogen peroxide oxidation is effective but limited by reactions with calcium hydroxide. After oxidation, the resulting arsenate waste is effectively stabilized using ferric sulfate.

4.6.2.3 Chlorine (Cl_2)

Chlorine is a good oxidant for As(III), but application must be made with caution in the treatment train avoiding any chance of formation of chlorine disinfection by-products.

4.6.2.4 Potassium permanganate

Potassium permanganate ($KMnO_4$) is one of the safest and most effective oxidants and costs much less than ozone or H_2O_2 while leaving practically no scope for harmful by-product formation.

Figure 4.23 illustrates how rejection of arsenic by nanofiltration membrane improves substantially with an increase of dose of $KMnO_4$. As the dose of $KMnO_4$ increases from 2 to 8 mg/L, rejection of arsenic by the most effective NF-1 membrane rises from 65% to over 98%.

Arsenic generally occurs in inorganic form and in two valence states— As(III) and As(V).While As(V) species dominate under aerobic or oxidizing conditions, As(III) species dominate under reducing conditions. As(III) species may be present as arseneous acid (H_3AsO_3) and arsenite ions ($H_2AsO_3^-$, $HAsO_3^{2-}$, AsO_3^3). As(V) exists as arsenic acid and arsenate ions. Therefore, the effects of an arsenic oxidation state can be eliminated by pre-oxidation of As(III) to As(V) where $KMnO_4$ is used as an oxidizing agent.

This oxidation rate equation using $KMnO_4$ as oxidant may be expressed as [20]

$$\frac{d}{dt}\left[As^{III}\right] = -K\left[As^{III}\right] \tag{4.12}$$

where K is the first order rate constant (s^{-1}). The process is first order with respect to As(III) and zero order with respect to $KMnO_4$ because $KMnO_4$ is present in excess.

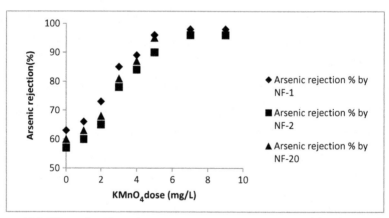

Figure 4.23 Effect of oxidant dose on percentage arsenic removal for ◆NF-1, ▪NF-2, and ▲NF-20. Pressure maintained at 12 kgf/cm². Arsenic concentration in initial feed water: 150 µg/L. Flow rate of water: 700 L/hr. Cross-flow velocity: 1.16 m/sec. Temperature: 35 °C [7].

Oxidation of arsenic in the presence of the oxidizing agent $KMnO_4$ takes place by following the reactions:

$$KMnO_4 \longrightarrow K^+ + MnO_4^-$$
$$H_2O \longrightarrow H^+ + OH^-$$
$$As^{3+} + MnO_4^- + 4H^- \longrightarrow As^{5+} + MnO_2 + 2H_2O$$

4.6.3 The treatment plant

An effective treatment scheme of oxidation–integrated nanofiltration is illustrated in Figure 4.24. The treatment scheme consists of a continuously stirred tank chemical reactor (CSTR) as the chemical preoxidation unit and a flat-sheet cross-flow nanofiltration module. The scheme is feed-and-bleed continuous type, where multiple passes of the arsenic-bearing water over the nanofiltration membrane surface is arranged. $KMnO_4$ as an oxidizing

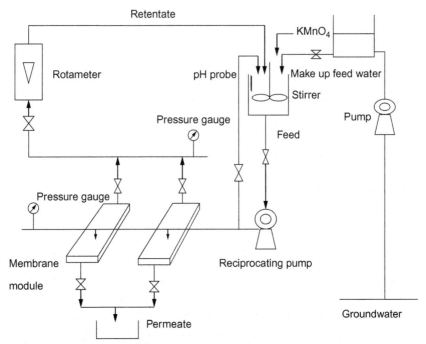

Figure 4.24 A hybrid treatment plant scheme integrating chemical oxidation with nanofiltration [13]. The circulating pump passes the feed water only a moderate pressure of 12–16 bar. A simple rotameter and the line valves can be used in monitoring and controlling flow of water. When concentration of arsenic in the reject stream exceeds a certain predetermined level, bleeding of the arsenic rejects is done.

agent for the trivalent arsenic is added continuously to the feed water tank based on the concentration of arsenic in groundwater.

4.6.4 Chemical oxidation-nanofiltration integrated with chemical stabilization

4.6.4.1 A novel complete system of arsenic separation and stabilization

A membrane-integrated hybrid treatment system has been developed for continuous removal of arsenic from contaminated groundwater with simultaneous stabilization of arsenic rejects for safe disposal (Figure 4.25) [14]. Both trivalent and pentavalent arsenic can be removed by cross-flow nanofiltration following a chemical preoxidation step for conversion of trivalent arsenic into pentavalent form. The very choice of the membrane module and its judicious integration with upstream oxidation and downstream stabilization results in continuous removal of more than 98% arsenic from water that contained around 190 mg/L of total suspended solid, 205 mg/L of total dissolved solid, 0.18 mg/L of arsenic, and 4.8 mg/L of iron at a pH of 7.2. The used flat-sheet cross-flow membrane module yields a high flux of around 145–150 L/(m^2 hr) at a transmembrane pressure of only 16 kgf/cm^2 without the need for frequent membrane replacement.

Transmembrane pressure, cross-flow rate through the membrane module, and oxidant dose were found to have pronounced effects on arsenic rejection and pure water flux. For the first time, an effective scheme for protection of the total environment has been ensured in this development where arsenic separated with high degree of efficiency has been stabilized in a solid matrix of iron and calcium under response surface optimized conditions. Continuous research on effective arsenic removal has culminated in a total and sustainable solution to the problem of arsenic contamination of groundwater by offering arsenic-free water at a reasonably low price of only 1.41$/m^3.

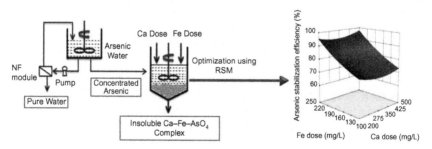

Figure 4.25 A total arsenic separation and stabilization scheme [14].

4.6.4.2 Transport and stabilization principles

Though arsenic is found both in organic and inorganic forms in water, the inorganic form dominates contaminated groundwater. Inorganic arsenic exists as either As(III) (H_3AsO_3) or As(V) (H_3AsO_4). As(III) behaves as a neutral molecule and it is poorly removed by most arsenic removal technologies. Hence, for waters containing As(III), preoxidation prior to treatment is required for efficient removal of arsenic. $KMnO_4$ can be used as an oxidizing agent for the preoxidation of As(III) to As(V).

Nanofiltration membranes separate components both by steric (size) and Donnan (electrical) exclusion mechanisms. The modified extended Nernst–Planck equation adequately describes separation of ionic or charged solute particles by such membranes and may be expressed as

$$J_{s,i} = (\alpha_i c_i \, V) - \left(D_i \frac{dc_i}{dx}\right) - \left(\frac{z_i \, c_i \, D_i \, F}{RT} \frac{d\psi}{dx}\right) \tag{4.13}$$

The flux (J_s) of ion i is the sum of the fluxes due to convection, diffusion, and electromigration. D_i is the diffusion coefficient of i through the membrane pores, which accounts for the friction of the components with the pore walls where α_i is the hindrance factor for convection and c_i is ionic concentration in feed. F, R, and T denote the Faraday constant, Universal gas constant, and temperature, respectively. z_i is the valence of the respective ions.

Again the solute flux can be expressed by

$$J_{s,i} = V \times C_{i,p} \tag{4.14}$$

where V is the solvent velocity. It may be expressed using a Hagen–Poiseuille type equation as:

$$V = \left(\frac{r_p^2 \, \Delta P_e}{8\eta\Delta x}\right) \tag{4.15}$$

r_p is the membrane pore radius, η is the viscosity of the solution, Δx is the membrane thickness, and ΔP_e is termed as effective pressure driving force, expressed as

$$\Delta P_e = dp = (\Delta p - \Delta \pi) = \left[\Delta p - \left\{RT\sum(C_{i,b} - C_{i,p})\right\}\right]$$

$\Delta \pi$ is the osmotic pressure difference and can be computed using bulk feed concentration ($C_{i,b}$) and permeate concentration ($C_{i,p}$) of the solute ions. The membrane surface concentration of solute ions can be measured by the Donnan equilibrium condition, which is a function of the Donnan potential of the membrane (ΔV_d) expressed as

$$c_i = (C_{i,b}\, \varphi_i) \exp\left(\frac{-z_i\, F \Delta V_d}{R\, T}\right) \qquad (4.16)$$

where Φ_i is the steric coefficient of ion i, z_i is the charge valence of ion i, F, R, and T are denoted as Faraday constant, universal gas constant, and temperature, respectively. V_d is the Donnan potential difference. This membrane surface concentration (c_i) of solute ions is used for the determination of permeates concentration, which is also used for the computation of rejection of the charge particles. The rejection (R_j) can be described by the following equation:

$$R_{i,j} = \left(1 - \frac{C_{i,p}}{C_{i,b}}\right) \qquad (4.17)$$

Arsenic can be precipitated out using varieties of coagulants. Ferric salts ($FeCl_3$ and $Fe_2(SO_4)_3$) can be used as coagulants leading to precipitation of ferric arsenate with higher insolubility than the calcium arsenate. Ferric ions, which are used as reducing agents, reduce the absolute values of the zeta potential of the particles leading to aggregation. Arsenic ions precipitate with the ferric ions on the coagulants, and thus increases the concentration of the coagulates as described below:

$$Fe^{3+} + H_3AsO_4 \longrightarrow FeAsO_4 \qquad (\text{solid precipitate})$$

Calcium is often used as a coagulant in the form of lime, hydrated lime, and calcium carbonate, leading to formation of largely insoluble calcium-arsenic compounds. The precipitation chemistry of arsenates and arsenite with hydrated lime ($Ca(OH)_2$) are well described by the following equations:

$$H_3AsO_4 + Ca^{2+} \longrightarrow Ca_3(AsO_4)_2 \qquad (\text{solid precipitate})$$
$$H_3AsO_3 + Ca^{2+} \longrightarrow CaHAsO_3$$

Addition of ferric ion to the arsenate compounds along with calcium ion produces $Ca–Fe–AsO_4$ compounds with a high degree of insolubility leading to the subsequent precipitation of the finally formed compound. The involved chemical reactions leading to the formation of $Ca–Fe–AsO_4$ compounds can be described as follows:

$$Fe^{3+} + H_3AsO_4 \longrightarrow FeAsO_4 \qquad\qquad \text{Precipitation}$$
$$Ca^{2+} + FeAsO_4 \longrightarrow Ca—Fe—AsO_4 \text{ complex} \quad \text{Coprecipitation}$$

4.6.4.3 Oxidation–nanofiltration–coagulation-integrated plant

This hybrid system consists of three basic units in sequence meant for pre-oxidation, nanofiltration, and stabilization. The total scheme of treatment from oxidation to stabilization is illustrated in Figure 4.26 [14].

4.6.4.4 Preoxidation unit

Preoxidation of trivalent arsenic to pentavalent form is done in a continuous stirred tank reactor (CSTR) prior to nanofiltration of the contaminated groundwater that contains both trivalent as well as pentavalent arsenic. $KMnO_4$ is used as the oxidizing agent and its optimum dose is found out experimentally for the used water characteristics. Stirring in the CSTR is done at 180 rpm. Figure 4.26 shows the schematic diagram of this preoxidation step as the first step of the membrane-integrated hybrid process.

4.6.4.5 Nanofiltration unit in flat-sheet cross-flow module

Nanofiltration is the second step in this hybrid scheme of treatment. Flat-sheet cross-flow membrane modules well known for their capability of providing long service without significant fouling are used for separation of arsenic in pentavalent form. Membrane fouling in this module is significantly lower than in the other widely used modules like spiral-wound, hollow fiber, plate and frame, or tubular types. Polyamide composite nanofiltration membrane such as NF-1 of Sepro Membranes Inc., USA, is found suitable in this module.

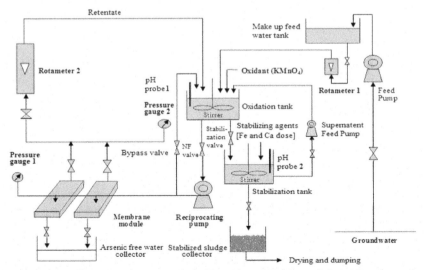

Figure 4.26 Schematic diagram of the membrane-integrated hybrid process for removal of arsenic from water and its stabilization in solid matrix [14].

Operating pressure is required to be maintained in the range of 12–16 kgf/cm^2. Cross-flow across the membrane module should be maintained in the range of 300–750 L/hr. The NF valve shown in Figure 4.26 is used to remove arsenic from contaminated groundwater while keeping "stabilization valve" in closed condition. After 5 to 6 months when stabilization is required, "stabilization valve" is opened to transfer concentrated arsenic-bearing water to stabilization tank, keeping the NF valve in closed position. Characterization of groundwater as presented in Table 4.5 may be done before and after hybrid treatment for assessing system performance.

4.6.5 Performance of the Nanofiltration Module

4.6.5.1 Pure water flux and rejection of arsenic and other contaminants

Figure 4.27 shows variation of pure water flux as well as arsenic rejection by the NF-1 nanofiltration membrane during cross-flow filtration run under varying transmembrane pressure.

The volumetric flux in LMH increases from 48 to 145 LMH as transmembrane pressure increases from 5 to 18 kgf/cm^2. Beyond a pressure of 16 kgf/cm^2, improvement in flux is very marginal. An operating pressure of 16 kgf/cm^2 may

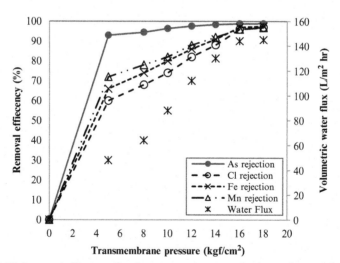

Figure 4.27 Removal efficency of various contaminants and volumetric water flux during nanofiltration under varying transmembrane pressure [14]. Experimental conditions: NF-1 membrane, preoxidized feed water with As concentration of 0.18 mg/L, Cl concentration 190 mg/L, Fe concentration 4.8 mg/L, Mn concentration 1.39 mg/L, pH 7.2, pressure range 5–18 kgf/cm^2, cross-flow rate 750 L/hr, cross-flow velocity 1.15 m/sec, temperature 308 K, preoxidized by KMnO$_4$.

thus be maintained as the optimum one. Figure 4.27 also indicates retention of arsenic and other contaminants (Cl, Mn, Fe) by nanofiltration membrane under varying applied pressure, and rejection shows a strong positive correlation with pressure. The removal efficiency of arsenic increases from 93% to more than 98% with increase of pressure from 5 to 18 kgf/cm^2 for a feed water initially containing around 0.20 mg/L arsenic. Chloride ions, manganese ions, and ferric ions get rejected 97%, 96.8%, and 97.5%, respectively, at the maximum pressure of 18 kgf/cm^2. In the solution-diffusion mechanism considered as the predominant transport mechanism of nanofiltration membrane, solute flux and solvent flux are uncoupled in nature and consequently, with the increase of applied pressure when solvent flux increases, it results in a commensurate increase in solute rejection [13,14].

4.6.5.2 Cross-Flow Rate and Flux Behaviour

Cross-flow rate has significant influence on pure water flux as illustrated [14] in Figure 4.28. Effects of such cross-flow on retention of charged solutes and volumetric water flux of pure water during nanofiltration have been well demonstrated [13]. Arsenic rejection is found to increase sharply from 91% to over 98% with an increase of cross-flow rate from 300 to 750 L/hr, whereas pure water flux rate increases from 78 to 144 LMH. Some other ions rejection also is affected by the cross-flow rate of the system.

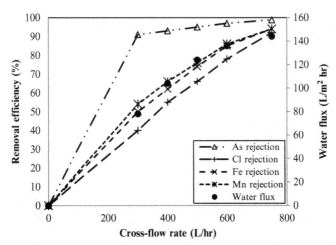

Figure 4.28 Arsenic removal and pure water flux during nanofiltration under varying cross-flow rates. Experimental conditions: NF-1 membrane, preoxidized feed water with As concentration of 0.18 mg/L, Cl concentration 190 mg/L, Fe concentration 4.8 mg/L, Mn concentration 1.39 mg/L, pH 7.2, pressure 16 kgf/cm^2, cross-flow rate range 300–750 L/hr, cross-flow velocity 1.15 m/sec, temperature 308 K, preoxidized by $KMnO_4$ [14].

Removal efficiency of chloride, manganese, and ferric ions increase from 56% to 97%, 60% to 97%, and 62% to 97.6%, respectively, as the cross-flow rate varies from 300 to 750 L/hr. Cross-flow rate plays a significant role in reducing concentration polarization by virtue of its sweeping action on the surface of the membrane preventing fouling of the membrane. Reduction in concentration polarization in turn increases effective charged surface area of the membrane, leading to an increase in arsenic separation from water where the Donnan exclusion principle dominates in transport mechanism. Reduced concentration polarization also enhances convective force, which in turn improves solvent flux due to the very uncoupling nature of the solvent and solute fluxes during nanofiltration. The overall drop in flux over a long 144 hr of operation is only 10–12%. Simple rinsing of the used membrabnes with 0.1 N NaOH and $10^{-?}$ M HNO_3 could remove the reversible fouling and restore flux almost to its original level.

4.6.5.3 Stabilization Unit

In the continuous run, contaminated water pumped from an underground aquifer is continuously fed to the reservoir tank at the same rate at which treated arsenic-free water is collected as filtrate. After months of continuous operation when concentration of contaminants such as Fe, As, Mn, Cl, TDS, and TSS reach high levels on the retentate side as presented in Table 4.7 the bleeding operation is done for purging concentrated sludge from the system for subsequent stabilization. The next downstream operation is thus stabilization of arsenic rejects in a solid matrix through the coagulation-precipitation process. The system provides for periodic withdrawal of arsenic rejects from the loop for precipitation and stabilization of arsenic.

4.6.5.4 Optimization of stabilization process

Response Surface Methodology (RSM) of Design Expert Software (Version 8.0.6) has been successfully used [14] in optimization of the parameters involved in the process of stabilization of arsenic in solid matrix through coagulation-precipitation. Initial concentration of arsenic in water, dose of ferric salt, dose of calcium salt, combinations of minerals, and pH have pronounced effects on stabilization of arsenic in a solid matrix. RSM can be used very effectively as a statistical tool for optimizing the process variables. RSM helps to find out the appropriate combination of the coagulant doses, initial concentration of arsenic, and pH. In the optimization process, important parameters like initial concentration of arsenic dose of calcium salt, dose of ferric salt, and pH may be considered independent variables whereas arsenic stabilization efficiency may be considered the dependent variable. The number of independent variables (v) is thus 4. Stabilization of arsenic here means removal or transfer of arsenic from the aqueous phase to the solid matrix.

The required number of experiments (N_{exp}) is calculated using the equation:

$$N_{exp} = (2^v + 2v + 6) = (2^4 + 2.4 + 6) = 30 \qquad (4.18)$$

As presented in Table 4.6, the five codes used in the RSM are $-\alpha$, -1, 0, $+1$, and $+\alpha$, indicating the minimum (−) and maximum (+) value of the variables; the values should be within this range. The following empirical quadratic polynomial explains the predicting values of the dependent variable Y (arsenic stabilization efficiency):

$$Y = \sigma_0 + \sum_{i=1}^{n} \sigma_i . x_i + \sum_{i=1}^{n} \sigma_{ii} . x_{ii}^2 + \sum \sigma_{ij} . x_i . x_j \qquad (4.19)$$

Y denotes the predicted response of arsenic stabilization, while σ_0, σ_i, σ_{ii}, and σ_{ij} are offset terms for the linear effects, square effects, and the interaction

Table 4.6 Range of the independent factors used in RSM

Independent Factors	Units	Symbol	Codded levels				
			$-\alpha$	-1	0	$+1$	$+\alpha$
As (V) concentration	mg L^{-1}	A	5	20	35	50	65
Ca dose	mg L^{-1}	B	50	200	350	500	650
Fe dose	mg L^{-1}	C	25	100	175	250	325
pH	–	D	3	5	7	9	11

Table 4.7 Characteristics of Arsenic-Contaminated Groundwater at Different Stages of Treatment

Water parameters	Unit	Feed side	Permeate side	Concentrated solution after 5 month operation
TSS	mg/L	196	BDL	29,100
TDS	mg/L	205	10	27,750
Conductivity	μs/cm	598	48	–
Salinity	–	0.45	0.03	–
pH	–	7.2	6.5	7.9
Chloride	mg/L	190	8.2	27,300
Manganese	mg/L	1.39	0.06	201
Iron	mg/L	4.80	0.15	490
Total arsenic	mg/L	0.18	0.00256	25
As(III)	mg/L	0.075	BDL	–
As(V)	mg/L	0.105	BDL	–

Source: Ref. [14].

Table 4.8 Optimum Conditions for Stabilization of Arsenic in Solid Matrix

Parameters	Unit	Optimum values
As(V) concentration	mg/L	25
Fe dose	mg/L	250
Ca dose	mg/L	500
pH	–	5
Weight of arsenic solid precipitate (for 5000 L)	g	38
Model predicted arsenic stabilization	%	98.0
Experimental arsenic stabilization	%	98.4

effects, respectively. x_i and x_j are the coded values of the independent variables. The required number of experiments under different operating conditions (Table 4.8) as suggested by Eq. (4.17) is conducted following a central composite design. The pH of the solution is adjusted using either concentrated HCl or 5 mm NaOH depending on the pH of the medium. Magnetic stirring at the rate of 140 rpm is done for 15 min in each case for fast dispersion of the reagents. After prolonged settling for 6–8 hr, the precipitate is collected by filtering the sample through filter paper (pore size 0.45 μm) and the filtrate is subsequently analyzed for residual arsenic concentration.

4.6.6 Assessment of stabilization

4.6.6.1 Leaching tests for assessing stabilization of arsenic rejects (Ca–Fe–AsO₄)

The solid arsenic mixture (7.5 mg) is taken for the TCLP (Toxicity Characterstics Leaching Procedure) test. During the TCLP test, an extraction fluid is prepared by mixing it with 5.7 mL glacial acetic acid and 64.3 mL of 1(N) sodium hydroxide in 1 L deionized water and a maintained pH of 5 (USEPA, 1992). The fresh solid precipitate is mixed with the extraction fluid (20 × the solid weight). The mixing slurry is agitated for 18 hr at room temperature (25 °C). After agitation the sample is given 6 to 8 hr for settling; after settling the extract is filtered through a filter paper of 0.45 μm pore size. The California Wet Extraction Test (CWET) [27] is also performed for assessing the stability of arsenic in the arsenic-bearing solid precipitate. The principle of this test is similar to that of TCLP, where the distinguishing features lie in using 0.2 M sodium citrate and an agitation time of 48 hr in the CWET.

4.6.6.2 Chemical analysis for assessing stabilization of arsenic rejects (Ca–Fe–AsO₄)

Concentration of arsenic in water is measured using an atomic absorption spectrophotometer. After the conversion of arsenic to volatile hydride the analysis is done using the flame-FIAS technique with a 193.7-nm wavelength [21]. Five percent (w/v) of potassium iodide (KI) and 5% (w/v) of ascorbic acid was used as a reducing agent for reducing As(V) to As(III). An oxy-acetylene flame was used to atomize the sample and the FIAS technique was used to inject the exact volume of the sample into the carrier system. As(III) is measured after suppressing the As(V) from the total arsenic by the adjusting pH 3 with the help of sodium hydroxide or citric acid. The subtraction of As(III) from the total arsenic gives the As(V) concentration. Removal of arsenic from aqueous medium is computed using the initial value (C_F) and the residual value (C_p) of the untreated and treated water, respectively, using the following equation:

$$R_j\,(\%) = \left(1 - \frac{C_P}{C_F}\right) \times 100 \qquad (4.20)$$

Concentration of manganese and iron present in groundwater are also measured using an atomic absorption spectrophotometer. The analysis is done with the flame-FIAS technique with a 403.1- and 372-nm wavelength for manganese and iron, respectively, and a maintained slit gap of 0.2 nm. The standards for both elements are prepared the following way: 1 g manganese metal is mixed with HNO_3 and diluted to 1 L with 1% (w/v) HCl and 1 g of iron salt in 50 mL of HNO_3 and diluted to 1 L. Nitrous oxide-acetylene flame is used to atomize the sample and the FIAS technique is used to inject the exact volume of the sample into the carrier system.

4.6.6.3 Structural Analysis

Fourier Transform Infrared (FT-IR) spectroscopy (viz. by Nexus 670, Thermo Electron Corporation, USA) is done for the stabilized arsenic precipitate following leaching tests. Arsenic-bearing precipitates (before and after leaching) are dried in an oven at 45 °C for 48 hr, and then analyzed by FT-IR spectroscopy. Forty mg of potassium bromide and 2 mg of finely ground sample is mixed well for preparing transparent pellets, for the determination of bonding characteristics. FT-IR analysis of arsenic-bearing precipitates before and after leaching further confirms stability of arsenic binding.

Figure 4.29 illustrate the FT-IR spectrum of the solid arsenic precipitate ($Ca–Fe–AsO_4$) as obtained [14] during arsenic stabilization before and after the standard leaching test (TCLP), respectively. The reactive functional groups like Fe–O, As–O, S–O, and OH stretching band that lie in the range of 500–3,500 cm^{-1} in solid precipitate are illustrated in Figure 4.29a. The water-stretching broad band as observed at 3,000–3,600 cm^{-1} wave number indicates the presence of crystalline hydrate in solid precipitate.

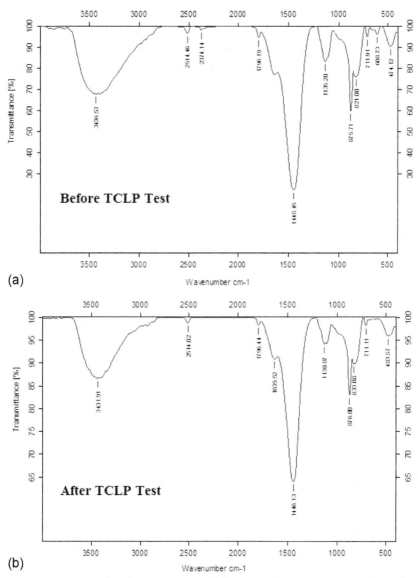

Figure 4.29 FTIR results of Ca–Fe–AsO$_4$ at different conditions. (a) FTIR results before leaching test; (b) FTIR results after leaching test (TCLP).

The water-calcium bonding is attributed to 1796 band. The peaks at 474, (821–875), 1,136 cm^{-1} wavenumber indicate the presence of Fe–O, As–O, S–O, and O–H stretching bands, respectively. Ca–O stretching band is found at 608 cm^{-1} wavenumber and it is not visible in Figure 4.29b, which suggests that some Ca^{2+} ions leached out from the solid. The overall FT-IR analysis shows (Figure 4.29a and b) that there is no significant change in peak wavenumber before and after leaching tests of the arsenic-bearing precipitates.

Stabilization of arsenic in the solid precipitate is found to be significantly influenced by arsenic concentration, ferric sulfate dose, calcium hydroxide dose, and pH of the medium as exhibited by Figure 4.30.

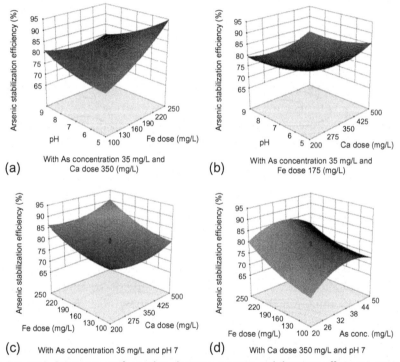

Figure 4.30 Response surface plot showing arsenic stabilization efficiency against operating parameters. (a) Effects of Fe concentration and pH on arsenic stabilization efficiency at arsenic concentration of 35 mg/L and Ca concentration 350 mg/L, process temperature 35 °C, stirring time 15 min, stabilization time 8–10 hr; (b) effects of Ca concentration and pH on arsenic stabilization efficiency at arsenic concentration of 35 mg/L and Fe concentration 175 mg/L; (c) effects of Fe concentration and Ca concentration on arsenic stabilization efficiency at arsenic concentration of 35 mg/L and pH 7; (d) effects of Fe concentration and initial arsenic concentration on arsenic stabilization efficiency at constant Ca concentration of 350 mg/L and pH 7 [14].

Figure 4.30a shows that at a fixed initial arsenic concentration of 35 mg/L and calcium ion dose of 350 mg/L, achieved stabilization efficiency of arsenic is as high as 95% at a pH value less than 9 and ferric salt dose of 250 mg/L.

Figure 4.30b describes the conjugate effects of pH and calcium hydroxide dose on arsenic stabilization. Surface plots in the figure show that the highest arsenic stabilization efficiency is achieved at the lowest pH (5) and the highest calcium dose of 500 mg/L while arsenic concentration and ferric sulfate dose are maintained at 35 and 175 mg/L, respectively.

Figure 4.30c shows that the arsenic stabilization efficiency increases from 86% to 92% with an increase of the calcium dose from 200 to 500 mg/L at a ferric ion dose of 250 mg/L. The stabilization efficiency decreases gradually from 92% to 78% for a decrease of the ferric ion dose from 250 to 100 mg/L.

Figure 4.30d shows the effects of ferric ion concentration and initial arsenic concentration on arsenic stabilization at an optimum calcium dose of 350 mg/L and pH of 7.0. The arsenic stabilization efficiency increases from 78% to 86% with increase of arsenic concentration up to 35 mg/L with a fixed coagulant dose of ferric salt (250 mg/L). Beyond the 35 mg/L of arsenic concentration, the efficiency starts to decrease due to tempting collision between the colloids. A typical set of optimum values of the process variables involved in arsenic stabilization are shown in Table 4.8 [14].

Use of a response surface optimization technique in arsenic stabilization can help in arriving at optimum conditions of stabilization, eliminating the mutual interaction effects of the parameters in earlier studies. This in turn leads to more effective stabilization of arsenic in a solid matrix. Such a scheme of removal of arsenic from contaminated water with simultaneous stabilization in a solid matrix establishes cross-flow nanofiltration of arsenic bearing preoxidized water as a continuous new process. Calculation shows that after 5 months of continuous operation, rejected ferric ion concentration in the feed tank reaches a level of 490 mg/L concentration. So an extra iron dose (250 mg/L) need not be added, which makes the process much more cost effective.

4.7 MODELING AND SIMULATION OF OXIDATION-NANOFILTRATION HYBRID PROCESS FOR SCALE UP

4.7.1 The principles

Nanofiltration models [13–19] in connection with ion separation basically exploit the equilibrium partitioning concept to describe the distribution of ions at the pore inlet and outlet through the extended Nernst–Planck approach. However, in nanofiltration modeling, mathematical complexity often stands in the way of easy computation. Models naturally vary with the solvents and solutes of interest as well as the types of membrane modules and fluid flow patterns. This necessitates system-based models that can be easily implemented with confidence.

Based on the present scheme of separation of arsenic from contaminated groundwater using a cross-flow nanofiltration membrane module coupled with chemical oxidation, a mathematical model has been developed [13] with the extended Nernst–Planck approach. For charged solutes the separation mechanism is the Donnan exclusion, which, compared to other pressure-driven membrane processes, has a pronounced effect on the separation performance of nanofiltration membranes. Due to the charged nature of the membrane, solutes with opposite charge compared to the membrane charge (counter-ions) are attracted, while solutes with a similar charge (co-ions) are repelled. At the membrane surface, a distribution of co- and counter-ions occurs, thereby effecting additional separation. Based on the simplified kinetics as described earlier and the principles of transport through NF membranes as described in the previous sections, model equations are formulated that involve some realistic assumptions in order to avoid computational complexity. For example, the NF membrane is assumed to consist of a bundle of identical straight cylindrical pores of radius r_p and length Δx (with $\Delta x >> r_p$). The effective membrane charge density (X_d) that plays the most critical role in the Donnan exclusion is also assumed to be constant throughout the membrane. While considering concentrations of the components and electrical potential within membrane pores, radially averaged values are counted. In view of the low concentration ranges of the involved solutes in clear groundwater, osmotic pressure difference ($\Delta \pi$) may be assumed to be insignificant. Salvation barrier energy in this context (ΔW_i) may be neglected. All trivalent arsenic present in water is converted to pentavalent form with prior oxidation. With this understanding, the overall integrated process may be mathematically captured through the following equations.

4.7.2 Model equations
4.7.2.1 Overall mass balance of aqueous solution in reactor unit

$$\rho_0 A\left(\frac{dh}{dt}\right) = F_i\rho_i + F_{ri}\rho_{ri} - F_0\rho_0 \tag{4.21}$$

where ρ_i, ρ_0 are densities (kg/m^3) of water at the inlet and outlet and ρ_r is the density of the oxidant. F_i, F_0 are the volumetric flow rates (m^3/sec) of the inlet and outlet sections and F_r is the oxidant volumetric flow rate. A is the cross-sectional area of the tank (m^2).

4.7.2.2 Mass balance of As(V) in reactor unit

$$V'C_a = V'C_{a0} + V'KC_{ai}t \qquad (4.22)$$

where C_a is the final As(V) concentration (µg/L), which acts as a charge particle concentration or feed concentration in nanofiltration; C_{ai} and C_{a0} are the initial As(III) and As(V) concentrations (µg/L) present in contaminated groundwater; K is the oxidant rate constant (3.2×10^{-3} s^{-1}); t is the time required to convert all As(III) to As(V). Eq. (4.14) may be recast as

$$C_a = C_{a0} + KC_{ai}t \qquad (4.23)$$

The flux for the charged particle through the nanofiltration (NF) membrane can be measured using the extended Nernst–Planck equation [27,28]. A modified Nernst–Planck equation determines ionic flux (Na$^+$) from the NaCl solution, which is passed through a nanofiltration membrane [29]. In the present investigation, transport of As(V) ions through NF membrane may be expressed as:

$$J_i = -D_{i,p}\frac{dc}{dx} - \frac{Z_i c_i D_{i,p} F}{RT}\frac{d\psi}{dx} + K_{i,c}c_i V \qquad (4.24)$$

The flux (J) of ion i is the sum of the fluxes due to convection, diffusion, and electromigration. $D_{i,a}$ is the bulk diffusion coefficient of component i; $D_{i,p}$ is the diffusion coefficient of i through the membrane pores, which accounts for the component friction with the pore walls; and $K_{i,d}$ and $K_{i,c}$ are the hindrance factors for diffusion and convection, respectively.

Solvent velocity through nanofiltration membrane may be expressed using the Hagen–Poiseuille type equation as

$$V = \frac{r_p^2 \Delta P_e}{8\eta \Delta x} \qquad (4.25)$$

where ΔP_e is termed as effective pressure driving force and expressed as $\Delta P_e = dp = (\Delta p - \Delta \pi)$. In that case $\Delta \pi$ is termed osmotic pressure difference.

The potential gradient through the membrane can be derived from the Extended Nernst–Planck equation (Eq. 4.16) and may be expressed as

$$\frac{d\psi}{dx} = \frac{\dfrac{z_1 V}{D_{1,p}}\left(K_{1,c}c_1 - C_{1,p}\right) + \dfrac{z_2 V}{D_{2,p}}\left(K_{2,c}c_2 - C_{2,p}\right)}{\dfrac{F}{RT}\left(z_1{}^2 c_1 + z_1{}^2 c_1\right)} \tag{4.26}$$

The electro neutrality conditions within the pore and the permeate solutions are

$$z_1 c_1 + z_2 c_2 + X_d = 0 \tag{4.27}$$

$$z_1 C_{1,p} + z_2 C_{2,p} = 0 \tag{4.28}$$

where X_d = membrane charge density, mol/m^3. From Eqs. (4.19) and (4.20), it may be shown that

$$c_1 = \frac{z_2 c_2 + X_d}{-z_1} \tag{4.29}$$

$$C_{1,p} = \frac{-z_2}{z_1} C_{2,p} \tag{4.30}$$

Substituting c_1 and $C_{1,p}$ (from Eqs. 4.21 and 4.22) into Eq. (4.18) yields [18]

$$\frac{F}{RT}\frac{d\psi}{dx} = \frac{\left(\dfrac{K_{1,c}V}{D_{1,p}} - \dfrac{K_{2,c}V}{D_{2,p}}\right)c_2 - \left(\dfrac{V}{D_{1,p}} - \dfrac{V}{D_{2,p}}\right)C_{2,p} - \left(\dfrac{K_{1,c}VX_d}{D_{1,p}}\right)}{2c_2 - X_d} \tag{4.31}$$

Following the principle of the Donnan equilibrium condition, we may calculate membrane wall concentration of both components as the following, where the Donnan equilibrium condition is expressed as

$$\frac{c_i}{C_i} = \varphi_i \exp\left(\frac{-z_i F \Delta \psi_d}{RT}\right)\exp\left(\frac{-\Delta W_i}{kT}\right) \tag{4.32}$$

where the solvation energy barrier is expressed as

$$\Delta W_i = \frac{z_i{}^2 e^2}{8\pi\varepsilon_0 a_i}\left(\frac{1}{\varepsilon_p} - \frac{1}{\varepsilon_b}\right) \tag{4.33}$$

and where k = the Boltzmann constant and the stokes radius (a_i) is expressed by

$$a_i = \frac{kT}{6\pi\eta D_{i,a}} \tag{4.34}$$

Neglecting the solvation energy barrier, the Donnan equilibrium may be expressed as

$$\frac{c_i}{C_i} = \varphi_i \exp\left(\frac{-z_i F \Delta \psi_d}{RT}\right) \tag{4.35}$$

Because the value of the diffusive coefficient ($D_{i,a}$) (Table 4.1) for ion 1 (H^+ ion) is much greater than ion 2 ($H_2ASO_4^-$ ion), the value of Stokes' radius for ion 1 is very close to zero.

The concentration gradient for ion 2 can be derived from the ENP equation (Eq. 4.16) and it can be expressed by:

$$\frac{dc_2}{dx} = \frac{V}{D_{2,p}}\left(K_{2,c}c_2 - C_{2,p}\right) - z_2 c_2 \frac{F}{RT}\frac{d\psi}{dx} \tag{4.36}$$

There is, of course, a corresponding expression for ion 1, but our aim is to develop the model for negatively charged ions (i.e., $H_2ASO_4^-$ ion), so it is not necessary to calculate for ion 1 (i.e., H^+ ion).

Substituting Eq.(4.21) into Eq. (4.23) and rearranging yields

$$\frac{dc_2}{dx} = \frac{\left(\dfrac{K_{1,c}V}{D_{1,p}} + \dfrac{K_{2,c}V}{D_{2,p}}\right)c_2[c_2 - X_d] - \left[c_2 C_{2,p}\left(\dfrac{V}{D_{1,p}} + \dfrac{V}{D_{2,p}}\right)\right] + \left(\dfrac{V C_{2,p} X_d}{D_{2,p}}\right)}{2c_2 - X_d} \tag{4.37}$$

This equation indicates that the order of the numerator is higher than the denominator. The concentration gradient will be effectively constant (and hence the concentration profiles linear) provided that the effect of the c_2 term is relatively small. Under these conditions, the concentration gradient can be approximated as follows:

$$\frac{\Delta c_2}{\Delta x} = \frac{\left(\dfrac{K_{1,c}V}{D_{1,p}} + \dfrac{K_{2,c}V}{D_{2,p}}\right)c_{2,av}[c_{2,av} - X_d] - \left[c_{2,av} C_{2,p}\left(\dfrac{V}{D_{1,p}} + \dfrac{V}{D_{2,p}}\right)\right] + \left(\dfrac{V C_{2,p} X_d}{D_{2,p}}\right)}{2c_{2,av} - X_d} \tag{4.38}$$

The Donnan potential at the pore inlet ($x=0$) is the same for both ions and it is obtained from Eq. (4.24).

$$\Delta \psi_d(0) = -\frac{RT}{F}\left[\ln\left(\frac{c_1(0)}{\varphi_1 C_f}\right)\right] = \frac{RT}{F}\left[\ln\left(\frac{c_2(0)}{\varphi_2 C_f}\right)\right] \tag{4.39}$$

Algebraic manipulation of Eq. (4.31) with Eq. (4.28) yields

$$c_2(0) = \frac{X_d + \sqrt{(X_d^2 + 4\varphi_1 \varphi_2 C_f^2)}}{2} \tag{4.40}$$

Similarly, an equivalent quadratic expression at the pore outlet ($x = \Delta x$) gives

$$c_2(\Delta x) = \frac{X_d + \sqrt{\left(X_d^2 + 4\varphi_1\varphi_2 C_{2,p}^2\right)}}{2} \tag{4.41}$$

Here $c_2(\Delta x)$ is calculated with a guess value of $C_{2,p}$ and the guess value is checked with a new $C_{2,p}$ value from Eq. (4.36).

$$\Delta c_2 = c_2(\Delta x) - c_2(0) \tag{4.42}$$

$$c_{2,av} = \frac{c_2(0) + c_2(\Delta x)}{2} \tag{4.43}$$

Rearrangement of Eq. (4.30) yields the following explicit expression for $C_{2,p}$:

$$C_{2,p} = \frac{(Pe_1 + Pe_2)X_d c_{2,av} - (Pe_1 + Pe_2)c_{2,av}^2 + (2c_{2,av} - X_d)\Delta c_2}{\left(\dfrac{Pe_2}{K_{2,c}}X_d\right) - \left(\dfrac{Pe_1}{K_{1,c}} + \dfrac{Pe_2}{K_{2,c}}\right)c_{2,av}} \tag{4.44}$$

where the Peclet number, Pe_i, is defined by the Hagen–Poiseuille definition, and is expressed as

$$Pe_i = \frac{K_{i,c}V\Delta x}{D_{i,p}} \tag{4.45}$$

Rejection is calculated by the following expression:

$$R_j = 1 - \frac{C_{2,p}}{C_f} \tag{4.46}$$

4.7.2.3 Computation Procedure

The model equations thus developed are solved using these steps:

1. Total arsenic concentration (assuming all As(III) converted to As(V)) C_f in the feed solution to the membrane separation unit is first computed from the mass balance equations (using Eq. 4.14).
2. This feed concentration value is used in the model equations to compute permeate concentration in terms of arsenic.
3. By using the values of arsenic concentration in the filtrate and the feed water, calculation for the total arsenic rejection is done and the variation of arsenic rejection is correlated with the variation of KMnO$_4$ doses.
4. Membrane surface concentrations of both ions (H$^+$ and H$_2$AsO$_4^-$) are computed using Eq. (4.24).

5. Eq. (4.32) is then used to calculate $c_2(0)$ with a known feed concentration (C_f) and Eq. (4.33) is used to find $c_2(\Delta x)$ with an assumed permeate concentration (C_p) value.
6. $C_{2,av}$ and Δc_2 are computed using Eqs. (4.35) and (4.34) and checking the guess value of C_p using Eq. (4.36).
7. Arsenic rejection is then calculated.
8. Pure water flux is calculated by Eq. (4.16) and solvent velocity (V) is calculated by the Hagen–Poiseuille equation.
9. An iterative method is adopted to compute rejection and flux using some assumed value of surface charge density (X_d) until the assumed value converges with the experimental value.

4.7.3 Determination of the model parameters

4.7.3.1 Computation of flow rate and concentration of oxidant

The flow rate of the oxidant is determined using a factor considering the stoichiometry of the reaction.

$$F_{ri} = f_1 \times F_i \text{ where } f_1 < 1$$

The density of the treated water at the outlet is determined considering the average density of feed raw water and the oxidant.

$$\rho_0 = \frac{F_i \rho_i + F_0 \rho_0}{F_i + F_0} \text{ and } F_0 = F_i + F_{ri}$$

Cross-sectional area and volume of the reactor are computed as

$$A = \frac{\pi D^2}{4} \text{ and } V' = h \times A$$

4.7.3.2 Computation of pore radius (r_p) and effective membrane thickness (Δx)

Membrane pore radius (r_p) and effective membrane thickness (Δx) are calculated from the model by the comparison of experimental data with model data from the separation of uncharged solutes (sucrose).

4.7.3.3 Hindered diffusivity is calculated by [12]

$$D_{i,p} = D_{i,a} \times K_{i,d}$$

where $K_{i,d} = \left(1.0 - 2.3\lambda_i + 1.154\lambda_i^2 + 0.224\lambda_i^3\right)$ and $\lambda_i = \dfrac{r_{i,s}}{r_p}$

4.7.3.4 Computation of Peclet number by the Hagen–Poiseuille equation

$$Pe_i = \frac{K_{i,c} V \Delta x}{D_{i,p}}$$

where $K_{i,c} = (2 - \varphi_i)(1.0 + 0.054\lambda_i - 0.998\lambda_i^2 + 0.44\lambda_i^3)$ and $\varphi_i = (1 - \lambda_i)^2$.

Through statistical analysis, where Willmott d-index and relative errors are computed considering the experimental findings with the model predictions, such a model is found to be very good [13]. A very good model performance is indicated if the Willmott d-index $d \geq 0.95$ and relative error $RE \leq 0.10$. In the graphical comparisons presented in Figures 4.31–4.35, the model predictions are found to corroborate very well with the experimental findings.

4.7.4 Governing parameters in preoxidation–nanofiltration process performance

4.7.4.1 Oxidation dose and arsenic rejection

Figure 4.31 illustrates how retention of arsenic by nanofiltration membrane depends on oxidation dose. Increased oxidation dose results in higher retention of arsenic. The trend is the same for all three NF membranes as illustrated in Figure 4.31. However, beyond a certain maximum dose of around

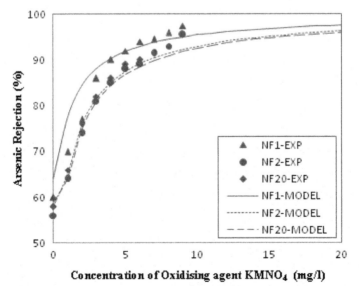

Figure 4.31 Variation of arsenic rejection with oxidation dose [13].

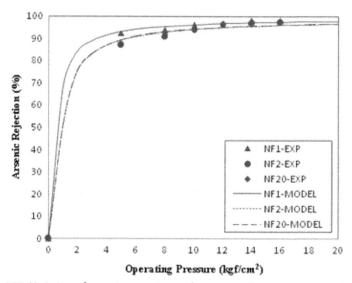

Figure 4.32 Variation of arsenic retention with transmembrane pressure [13].

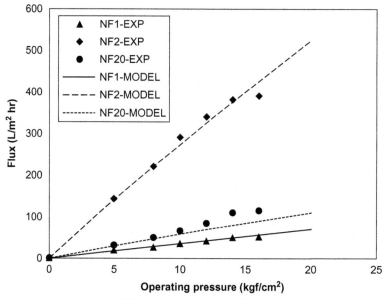

Figure 4.33 Flux behavior of NF membrane in cross-flow mode [13].

8 mg/L of KMnO$_4$ no further improvement in arsenic retention is observed for a water sample having a defined initial arsenic concentration [13]. At the optimum concentration of the oxidizing agent, rejection of arsenic crosses 98.5%. This point thus indicates the total conversion of trivalent arsenic to

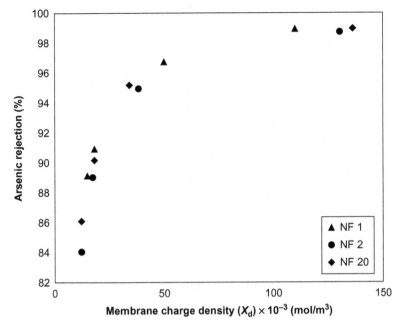

Figure 4.34 Effect of membrane charge density on arsenic rejection [13].

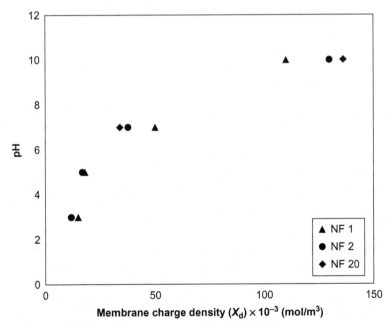

Figure 4.35 Effect of medium pH on membrane charge density.

pentavalent form. Obviously the optimum dose of the oxidizing agent will depend on the initial arsenic concentration as well as distribution of arsenic into different forms.

Model predictions are found to corroborate very well with the experimental findings, however, with some deviations at the lower levels of the oxidation dose. At higher levels of oxidation dose, such deviations are almost smoothed out. Such improvement in arsenic rejection with increase of oxidation dose could well be attributed to conversion of trivalent arsenic to pentavalent form. Arsenic in trivalent form basically remains in neutral form and escapes the impact of the dominating Donnan exclusion mechanism of nanofiltration separation. Preoxidation of As(III) by $KMnO_4$ results in substantial improvement in removal efficiency of the membranes when this trivalent arsenic gets converted to pentavalent form, which remains primarily in negatively charged form and experiences charge repulsion while passing over negatively charged nanofiltration membranes. Possibly all the arsenic present can get oxidized within a $KMnO_4$ dose of 8 mg/L and that may be the reason why no further improvement in rejection is noticed beyond a dose of 8 mg/L.

4.7.4.2 Transmembrane pressure and rejection

Figure 4.32 illustrates how retention of arsenic by nanofiltration membrane changes with transmembrane pressure.

Since operating pressure varies from 5 to 16 kgf/cm^2, a steady increase in arsenic rejection is observed for the three membranes up to a pressure of 14 kgf/cm^2. However, beyond the operating pressure of 14 kgf/cm^2, no further improvement in arsenic separation is observed except for a marginal increase in flux under the investigated maximum pressure range. So an operating pressure of 14 kgf/cm^2 may be taken as the optimum pressure in this case. With NF-1 membrane the highest rejection is around 98% followed by NF-20 (96.7%) and NF-2 (95.8%).

The rejection order followed by the three membranes is NF-1 > NF-20 > NF-2 at any particular applied pressure, and the trend corroborates with the experimental results. Upon fitting the data set to the linearized model in order, the effective membrane charge density X_d is obtained. Two transport mechanisms may work in the case of NF membranes. In solution diffusion mechanism, solute flux and solvent flux are uncoupled and as a result, an increase of solvent flux occurs with the increase of transmembrane pressure without increasing solute flux. That means solutes rejection will increase with the increase in transmembrane pressure. This explains the observed trend in arsenic rejection here under varying pressure.

In the size exclusion mechanism, relative sizes of the membrane pore and solute dimension assume an important role in determining degree of separation. When arsenic is present in water in both trivalent and pentavalent forms, size exclusion may play a significant role and therefore, the dominant mechanism in effecting separation here will largely be dependent on the forms in which arsenic is present in water. That is the reason why a low degree of arsenic separation is achieved by nanofiltration without preoxidation. When As(III) is totally converted to anionic As(V), charge repulsion becomes dominant in solute transport. The solution diffusion mechanism sets in and the uncoupled nature of solute flux and solvent flux explains why rejection increases with pressure. Model predictions on rejection here quite satisfactorily corroborate with the experimental observations. Such a degree of correlation of the model predictions with the experimental data have been elucidated in the error analysis section.

4.7.4.3 Transmembrane pressure and water flux

Permeate flux data from experimental investigations as well as from the model presented in Figure 4.33 shows that the permeate flux increases with the increase of transmembrane pressure for all the NF membranes and it was found that permeate flux varied linearly with applied pressure. Such flux behavior of membranes with transmembrane pressure during nanofiltration is well established in the literature.

However, investigations here throw light on relative flux enhancement with applied pressure and help select the best membrane with consideration of flux as well as rejection behavior of the membranes. NF-2 membrane yielded the highest flux due to its large pore radius (0.57 nm) whereas the NF-1 membrane (0.53 nm) yielded the lowest flux due to its small pore radius. NF-2 exhibited the highest degree of flux increase with increase of operating pressure against a relatively small extent of increase for NF-1. However, at 14 kg/cm^2 pressure, the flux of the NF-1 membrane of around 50–51 L/m^2.h is reasonably good and could be acceptable for scale up. The NF-20 membrane yields a flux intermediate between the two former types since its pore radius (0.54 nm) was in between the others. The developed model predicts such an increase of flux following an increase in operating pressure, and the experimental findings agree well with such predictions.

4.7.4.4 Membrane charge density and arsenic rejection during nanofiltration

Figure 4.34 illustrates variations of arsenic rejection by NF membranes with membrane charge density. Higher rejection is observed with higher membrane charge density, which is basically a negative charge on the surface of the composite polyamide nanofiltration membrane. As charge repulsion increases retention of arsenic also increases proportionally.

It is found [13] that with an increase of pH of arsenic-contaminated groundwater from 3 to 10, arsenic rejection increases to 99% with preoxidation of groundwater. Up to a pH value of 8.0, As(III) remains largely as a neutral molecule while As(V) remains as an anion. With change of pH, As(V) speciation even changes from monovalent ($H_2AsO_4^-$) to divalent form $HAsO_2^{2-}$ enhancing further retention of arsenic due to the Donnan exclusion. pH, membrane charge density, and arsenic rejection are well correlated. Figure 4.35 shows the effect of pH on membrane charge density. It is observed that pH has a strong positive correlation with membrane charge density.

The NF-1 membrane has the highest membrane charge density whereas NF-20 has the lowest value. This explains again the highest arsenic rejection by the NF-1 membrane followed by NF-2 and NF-20, resulting from charge repulsion or Donnan exclusion. Since the value of r_p for the three membranes was quite large compared to the effective ion radius (Table 4.1) of H^+ and $H_2AsO_4^-$ (for H^+, $r_s = 0.025$ nm and for $H_2AsO_4^-$, $r_s = 0.258$ nm [20]) the differences in the rejection of arsenic in these cases basically are due to the differences in X_d value. The rejection of arsenite and arsenate is affected by electrical charge and charge valence (as discussed in the theoretical section). Thus pH has dual effects on separation—one through change of arsenic speciation and the other through change of membrane charge density. Figure 4.35 reflects such variation of arsenic rejection with effective charge membrane density (X_d) determined from the linearized model.

4.8 OPTIMIZATION AND CONTROL OF MEMBRANE-BASED PLANT OPERATIONS

4.8.1 ARRPA: The optimization and control software

A simulation software (ARRPA) has been developed [30] in Microsoft Visual Basic for optimization and control of the novel membrane-integrated arsenic separation plant described in the previous section. The software is included in this book and the corresponding CD is attached for the readers. Using the same model from Eqs. (4.21)–(4.46) the user-friendly, menu-driven software based on a dynamic linearized mathematical model has been developed for the hybrid treatment scheme. The model captures the chemical kinetics in the pretreating chemical reactor and the separation and transport phenomena involved in nanofiltration. The software successfully predicts performance of the oxidation-nanofiltration hybrid treatment plant described in the previous section. High values of the overall correlation coefficient ($R^2 = 0.989$) and Willmott d-index (0.989) are indicators of the capability of the software in analyzing performance of the plant. The software permits preanalysis and manipulation of input data, helps in optimization, and exhibits performance of an integrated plant visually on a graphical platform. Performance analysis of the whole system as well as the individual units is possible using the tool. The software is the first of its kind in its domain, and in the well-known Microsoft Excel environment it is likely to be very useful in the successful design, optimization, and operation of an advanced hybrid treatment plant for removal of arsenic from contaminated groundwater.

Membrane-integrated treatment plants promise a high degree of purification and also high flux if run in an appropriate module. Monitoring and control of such plants are also very essential, and this is where the Visual Basic software tool steps in. The tool permits optimization of the operating variables and rapid analysis of performance of the constituent process units as well as the integrated system in a familiar Microsoft environment.

4.8.2 Software Description
4.8.2.1 Data sheet design

The arsenic separation plant analysis software is designed in Visual Basic 10. The software produces the visual graphics using the output values (water flux and arsenic rejection). The start-up page, or **General page**, is designed and designated as **Tool bar** and **Property box** options contained in the software as illustrated in Figure 4.36.

It consists of four different units such as **General, Reactor, Nanofiltration-1, Nanofiltration-2,** and **Nanofiltration-20** unit and contains several setting modes of application as well as several types of labeling such as **Software start-up, Screen Placement, Mode of Simulation, Save Settings, Mode of Data Handling,** and **Last up-gradation.**

Under the **Software start-up** option two tabs are found, **Data,** where all the data related to the process are stored, and **Show Helps,** which can be chosen to get proper guidance about the concerned operation using the software. The **Screen Placement** option permits visualization of different windows in different classes like **Tile Cascade, Tile Horizontally, Tile Vertically,** and **Arranged by Icons. Mode of Data Handling** permits

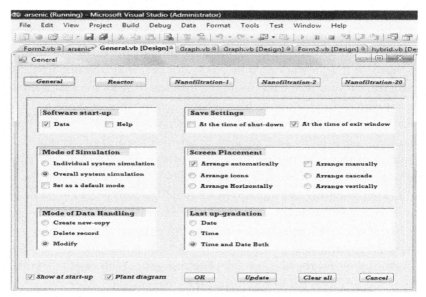

Figure 4.36 General settings window interface of the software.

the user to create and define the database for individual units and stores the values entered by the user. **Save Settings** is used to save the data and operation mode at the shutting down position of the computer windows. **Simulation Mode** optimizes the process pathway, allowing performance analysis following the individual or overall path. The working time can be stored and reused by the **Date of Working** option. **Software start-up**, **Ok**, **Update**, **Clear All**, and **Cancel** are the five options provided on every page of the data sheet. **Show at start-up** appears on each data sheet in the **Check Box** format and shows the data sheet at the start of the software use. **Ok** allows further processing and generates the results as real numbers as well as in graphical mode.

By choosing the **Update** tab, newly created data can be saved, whereby clicking the **Clear All** and **Cancel** tabs, a new data sheet can be rewritten by entirely erasing the existing data and cancelling the updated data, respectively. A schematic diagram of the whole treatment plant could be retrieved by clicking the **Plant Diagram** tab at the start-up page. After designing all the operational pages of the software, the designing program of this software can be written in the **Coding Page** with the help of model equations, which are accessed from the **View Code** option by right-clicking the user interface.

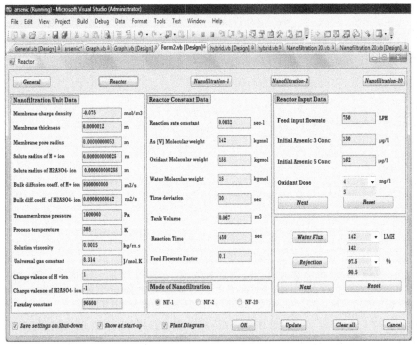

Figure 4.37 Input data sheet of reactor unit: effects of oxidant dose on arsenic removal.

Some of the typical input and output data sheets are presented here for understanding the software.

Figure 4.37 shows the effects of reactor input parameters like oxidant dose in the reactor tank of the hybrid system during arsenic removal. Water flux does not get affected much by the addition of an oxidizing agent. Using all the constant parameter values, arsenic rejection and water flux are predicted by the model equations, which are written in the **Coding Page**. The input value is inserted into the box by the **Text** option, which is positioned in **Property box**. Every text value is named in a short form and is used during coding. The input parameters of the reactor have been written in a **Combo box** option, which is changed by clicking the **Next** option. Clicking the **Water Flux** tab in the page yields the permeate water flow rate in L/(m^2 hr) while the **Rejection** tab shows the removal efficiency of arsenic. The resulting value of water flux and rejection appear in a **Combo Box** tab, where all the values for the different input parameters are stored. All the values are shown by clicking **Next**, and the **Reset** button is used to reset the values to their last working values.

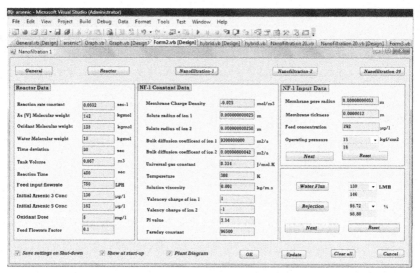

Figure 4.38 Input data sheet of the nanofiltration unit of NF-1 membrane: Effects of transmembrane pressure on water flux and arsenic rejection with initial arsenic concentration 292 μg/L, cross-flow rate 750 LPH, and water pH 9 [30].

Figure 4.38 shows the effects of input parameters during the transport of feed water through the NF membrane. Pore radius of the membrane, transmembrane pressure, overall arsenic concentration, and membrane thickness of the NF membrane are the input parameters in the software because of their effects on water flux and rejection of arsenic. Increase in transmembrane pressure from 1 to 20 kgf/cm^2 resulted in an increase of arsenic rejection from 87 to 99% and water flux from 10 to 181 LMH. Such rejection behavior of the NF membrane may be attributed to two transport mechanisms. One transport mechanisms of a NF membrane is a solution–diffusion mechanism, where the solvent flux and the solute flux are uncoupled. Thus an increase of solvent flux does not essentially lead to a rise in solute flux; rather it leads to its fall.

Figure 4.39a illustrates the effects of oxidation dose on percentage removal of arsenic. As discussed in the theoretical section, due to the role of the Donnan exclusion principle, conversion of arsenic from trivalent to pentavalent with the use of $KMnO_4$ as oxidant results in larger rejection and separation of arsenic from water during NF.

Figure 4.39b describes the effects of transmembrane pressure on arsenic removal and the production of pure water as filtrate of NF-1 membrane. The volumetric flux in LMH is calculated by the model equation and the same increases from 10 to 180 LMH as transmembrane pressure increases from 5 to 20 kgf/cm^2.

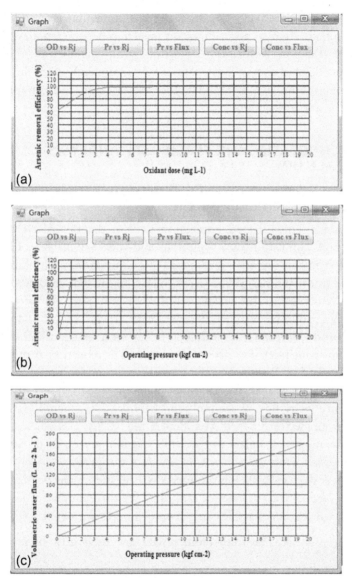

Figure 4.39 Output graphics of the model predictive data: (a) Effects of oxidant dose on arsenic removal efficiency of NF-1 membrane; (b) Effects of operating pressure on arsenic removal efficiency of NF-1 membrane; (c) Effects of operating pressure on volumetric water flux of NF-1 membrane [30].

Figure 4.39c illustrates the arsenic retention behavior of the nanofiltration membrane under varying applied pressure and it is found that the rejection has a strong positive correlation with pressure. The model-predicted arsenic removal efficiency increases from 87% to over 98% with an increase of pressure from 1 to 20 kgf/cm^2 for feed water containing around 180 μg/L arsenic initially. In a solution-diffusion mechanism considered as the predominant transport mechanism of nanofiltration membrane, solute flux and solvent flux are uncoupled in nature and consequently, with the increase of applied pressure when solvent flux increases, it results in a commensurate increase in solute rejection.

4.9 TECHNOECONOMIC FEASIBILITY ANALYSIS FOR SCALE UP

4.9.1 Membrane-integrated hybrid treatment plant: oxidation-nanofiltration system

The largest arsenic-affected population lives in the Bengal-Delta Basin. The average population of a typical arsenic-affected village in India and Bangladesh in this region is around 2,000. Considering a drinking water requirement of 10 L per day per head, such a village will need 20,000 liters (7,300 m^3) of arsenic-free water for drinking purposes. This section thus provides an economic analysis of a membrane-integrated hybrid treatment plant of production capacity of 20,000 L/day for a village in West Bengal, India considering price standards of the region. The oxidation-nanofiltration treatment plant as described in the previous sections can produce 98–99% pure water, and each flat-sheet cross-flow membrane module can typically yield a pure water flux of around 145 LMH if a NF-1 nanofiltration flat-sheet membrane from Sepro Membranes Inc., USA, is used. In a village setup, operating such a plant in two shifts spanning over a total 16 hr seems most practical. Thus each m^2 of the module can produce 16×145 L (or 2,320 L) per day. So 2,300 L is produced by 1 m^2 membrane per day. If each module possesses 0.25 m^2 surface area, then the total number of such modules that will be necessary for the desired production of 20,000 L/day is $\dfrac{20,000}{2,320 \times 0.25} = 35$. The sixth-tenth power law used [31] for scale up is defined as follows:

$$\text{Scale-up cost} = \text{Lab scale cost} \times \left(\frac{\text{Scale-up data}}{\text{Lab scale data}} \right)^{6/10}$$

Cost involvement in terms of capital and operation cost is calculated and presented in Table 4.9.

The cost assessment is based on the annualized investment cost and annualized operational cost. Annualized capital cost is computed by the following relationship:

Table 4.9 Cost of a 20,000 L/Day Capacity Oxidation–Nanofiltration Water Treatment Plant

Cost parameters *Capital cost*	No of equipment with specification	Cost value ($) *Cost ($)*
Cost for civil infrastructure	30 m^2 (10 m × 3 m) space	4,600
Membrane module cost	35 no of module (0.25 m^2 area)	7,600
Large volume tank cost	2 (fiber tank, 20,000 L capacity)	1,200
High flow pump	1 (submersible pump)	300
High pressure pump cost	1 (diaphragm pump, max. pr. 50 bar)	1,200
Cost for main feed pipe	60 m long and 0.5 m dia.	1,500
Others pipe fittings and	Rotameter (2), pr. gauge (2), pH probe (2)	1,400
Electrotechnical cost	Stirring motor (1)	
Total cost		**17,200**
Operating cost		*Cost ($/year)*
Electricity cost	Power consumption:1000 kwh/month	1,000
Membrane cost	Membrane needed: 9 m^2, Cost: 80 $/m^2	1,440
Membrane life: 6 months		
Labor cost	No of labor: 2 (334 $/month per head)	8,000
Total cost		**10,440**

$$\text{Annualized investment cost} = \left(\frac{\text{Total investment} \times \text{Cost recovery factor}}{\text{Year wise water production}}\right)$$

The cost recovery factor is dependent on plant project life ($n = 15$ years) and interest ($i = 9\%$) and it can be calculated by the following equation:

$$\text{Cost recovery factor} = \left(i(1+i)^n / (1+i)^{n-1} - 1\right)$$

Again, annualized operational cost can be computed by the following equation:

$$\text{Annualized operational cost} = \left(\frac{\text{Total operational cost}}{\text{Year wise water production}}\right)$$

$$\text{Annualized investment cost} = \left(\frac{\text{Total investment} \times \text{Cost recovery factor}}{\text{Year wise water production}}\right)$$

$$= \left(\frac{17,200 \times 0.13}{7,300}\right) = 0.3\,\$/\text{m}^3$$

$$\text{Annualized operational cost} = \left(\frac{\text{Total operational cost}}{\text{Year wise water production}}\right)$$

$$= \left(\frac{10,440}{7,300}\right) = 1.43\,\$/\text{m}^3$$

Thus annualized cost for production of 1 m³ drinking water is the summation of annualized investment cost and annualized operational cost = 0.3 + 1.43 = 1.73 $

$$\text{Annualized investment cost} = \left(\frac{\text{Total investment} \times \text{Cost recovery factor}}{\text{Year wise water production}}\right)$$

$$= \left(\frac{18,100 \times 0.13}{7,300}\right) = 0.33\,\$/\text{m}^3$$

$$\text{Annualized operational cost} = \left(\frac{\text{Total operational cost}}{\text{Year wise water production}}\right)$$

$$= \left(\frac{10,960}{7,300}\right) = 1.51\,\$/\text{m}^3$$

Thus annualized cost for production of 1 m³ drinking water is the summation of annualized investment cost and annualized operational cost = 0.33 + 1.51 = 1.84 $.

This may be rounded to 2 $/m³ or 0.002 $/L considering other miscellaneous costs, which is against the prevailing market price of 0.25 $/L of arsenic-free safe bottled drinking water currently being marketed in India.

The annualized cost as reflected in Table 4.10 for producing arsenic-free water seems quite affordable by the affected people. The dynamic mathematical model described here for a membrane-integrated hybrid system of arsenic separation from contaminated groundwater is quite successful in predicting the real plant performance as reflected in Table 4.11.

Economic analysis shows that the annualized cost in production of arsenic-free water is reasonably low, indicating the possibility of accepting the technology readily. The modeling and simulation study along with economic evaluation is expected to pave the way for scale up and design of arsenic separation plants using the membrane-integrated hybrid treatment scheme.

Table 4.10 Cost of a 20,000 L/Day Capacity Oxidation–Nanofiltration Stabilization Water Treatment Plant

Cost parameters	No of equipment with specification	Cost value ($)
Capital cost		*Cost ($)*
Cost for civil infrastructure	30 m² (10 m × 3 m) space	4,600
Membrane module cost	35 no of module (0.25 m² area)	7,600
Large volume tank cost	1 (fiber tank, 20,000 L capacity)	600
Low volume tank cost	1 (fiber tank, 5000 L capacity)	200
High flow pump	1 (submersible pump)	300
High pressure pump cost	1 (diaphragm pump, max. pr. 50 bar)	1,200
Low pressure pump cost	1 (water discharge pump)	300
Cost for main feed pipe	6 m long and 0.15 m dia.	1,800
Others pipe fittings and	Rotameter (2), pr. gauge (2), pH probe (2)	1,500
Electrotechnical cost	Stirring motor (2)	
Total cost		**18,100**
Operating cost		*Cost ($/year)*
Electricity cost	Power consumption:1200 kwh/month	1,200
Membrane cost	Membrane needed: 2.5 m², cost: 50 $/m²	1,440
Membrane life: 6 months		
Chemical cost	Calcium chloride (20 kg/year, cost: 10 $/kg)	200
	Ferric sulfate (10 kg/year, cost: 12 $/kg)	120
Labor cost	No of labor: 2 (334 $/month per head)	8,000
Total cost		**10,960**

Table 4.11 Statistical Error Analysis for Model Fitness Using Arsenic Rejection Data

Membrane	Relative error (RE)	Willmott index (d)	Model performance
NF-1	0.0032	0.980	Very good
NF-2	0.0020	0.958	Very good
NF-20	0.00282	0.966	Very good

4.10 CONCLUSION

The preoxidation–nanofiltration–coagulation stabilization plant for treatment of arsenic-contaminated groundwater is the first of its kind, offering a sustainable solution to a longstanding arsenic contamination problem of groundwater affecting millions of people across the world. While the preoxidation step followed by flat-sheet cross-flow nanofiltration ensures more than 98% removal of arsenic, the stabilization step offers a safe disposal route for the arsenic rejects. The novelty and beauty of the membrane-integrated hybrid scheme lies in its high degree of separation and stabilization efficiency in a very simple treatment scheme that promises 1000 L of safe drinking water at a price of only around $2.

Fouling is often considered a major hindrance in succesful long-term operation of a membrane-based plant. The new scheme uses a flat-sheet cross-flow membrane module that could largely overcome this problem by virtue of the sweeping action of the fluid flow itself. Low flux often is considered another stumbiling block that has also been overcome here as evident in quite high flux of around 140–150 LMH. The development of this hybrid treatment system in a novel scheme promises a total solution of the problem of ensuring a supply of safe drinking water to the millions of affected people in the arsenic-prone areas of the world.

NOMENCLATURE

c_i	concentration in membrane of ion i (mol/m^3)
$c_{i,av}$	average concentration of ion i (mol/m^3)
$C_{i,p}$	concentration in permeate of ion i (mol/m^3)
C_f	feed concentration (mol/m^3)
$D_{i,p}$	hindered diffusivity of ion i (m^2/s)
$D_{i,a}$	bulk diffusivity of ion i (m^2/s)
$K_{i,c}$	hindrance factor for convection of ion i
$K_{i,d}$	hindrance factor for diffusion of ion i
J_s	uncharged solute flux (mol/(m^2 sec))
$J_{i,s}$	ion flux (mol/(m^2 sec))
J_v	volumetric flux (m^3/(m^2 sec))
$J_{i,v}$	volumetric flux of ion i (m^3/(m^2 sec))
r_p	effective pore radius (nm)
r_s	uncharged solute radius (nm)
$r_{i,s}$	solute radius of ion i (nm)
R_j	rejection (%)
Δx	effective membrane thickness (m)
A_k	porosity of the membrane

X_d	effective charge membrane density (mol/m^3)
z_i	valence of ion i
F	Faraday constant
R	universal gas constant (J/(mol K))
T	absolute temperature (K)
η	dynamic viscosity of the solution (kg/(m sec))
ΔP	applied pressure difference (kgf/cm^2)
ΔP_e	effective pressure difference (kgf/cm^2)
Φ	steric coefficient
λ_i	ratio of solute radius to pore radius of ion i
a_i	stoke radius (m)
ε_b	bulk dielectric constant, dimensionless
ε_p	pore dielectric constant, dimensionless
ε_0	permittivity of free space (8.854×10^{-12} J^{-1} C^2 m^{-1})
$\Delta\psi_d$	Donnan potential difference (V)
k	Boltzmann constant (1.38066×10^{-23} J/K)
Pe_i	Peclet number of ion i, dimensionless
d	thickness of oriented solvent layer (m)
e	electronic charge (1.602177×10^{-19} C)
A	cross sectional area of the tank (m)
V	volume of the tank (m^3)
h	height of the tank (m)
c_{in}	inlet concentration of the feed water (mol/m^3)
c_{out}	outlet concentration of the feed water (mol/m^3)
$c_{m,i}$	membrane wall concentration of ion i (mol/m^3)
$c_{mi,av}$	average membrane wall concentration of ion i (mol/m^3)
$C_{pm,i}$	solute concentration in the permeate side of ion i (mol/m^3)
$C_{f,i}$	solute concentration in the feed side of ion i (mol/m^3)
$D_{c,i}$	diffusion coefficient of i (m^2/sec)
$D_{b,i}$	bulk diffusion coefficient of i (m^2/sec)
F_{in}	inlet flow rate (L/hr)
F_{out}	outlet flow rate (L/hr)
$H_{c,i}$	convective hindrance factor of ion i
$H_{d,i}$	diffusive hindrance factor of ion i
$J_{s,i}$	uncharged solute flux of ion i (mol/m^2 sec)
J_v	volumetric flux (m^3/m^2 sec)
$J_{i,v}$	volumetric flux of ion i (m^3/m^2 sec)
r_p	membrane pore radius (nm)
$r_{i,s}$	solute radius of ion i (nm)
$R_{j,As}$	overall arsenic rejection (%)
Δx_m	effectual membrane thickness (m)
A_k	membrane porosity
X_m	membrane charge density (mol/m^3)
T	process temperature (K)
ΔP_T	transmembrane pressure difference (kgf/cm^2)
ΔP	original pressure difference (kgf/cm^2)
A	cross-sectional area of the tank (m)
V_r	volume of the reactor (m^3)

h	height of the tank (m)
d	Willmott d-index
LMH	liter per meter square hour
LPH	liter per hour
RE	relative error
η_w	water viscosity (kg/m sec)
$\Delta\psi_p$	Donnan potential difference in volt
Φ	steric coefficient
λ_i	ratio of solute radius to pore radius of ion i
ANOVA	analysis of variance
BDL	below detected level
CWET	California wet extraction test
LMH	liter per meter square hour
RSM	response surface methodology
TCLP	toxicity characterstics leaching procedure
TDS	total dissolved solid
TSS	total suspended solid

REFERENCES

[1] Brandhuber P, Amy G. Alternative methods for membrane filtration of arsenic from drinking water. Desalination 1998;117:1–10.

[2] Lin T, Wu J. Adsorption of arsenite and arsenate within activated alumina grains: equilibrium and kinetics. Water Res 2001;35(8):2049–57.

[3] Lin CF, Wu CH, Lai HT. Dissolved organic matter and arsenic removal with coupled chitosan/UF operation. Sep Purif Technol 2008;60:292–8.

[4] Hering JG, Elimelech M. Arsenic removal by ferric chloride. J Am Water Works Assoc 1996;88(4):155–67.

[5] Han B, Runnells T, Zimbron J, Wickramasinghe R. Arsenic removal from drinking water by flocculation and microfiltration. Desalination 2002;145:293–8.

[6] Hsieh LC, Weng Y, Huang CP, Li KC. Removal of arsenic from groundwater by electro-ultra filtration. Desalination 2008;234:402–8.

[7] Sen M, Manna A, Pal P. Removal of arsenic from contaminated groundwater by membrane-integrated hybrid treatment system. J Membr Sci 2010;354:108–13.

[8] Vrjenhoek EM, Waypa JJ. Effect of operating conditions in removal of arsenic from water by nanofiltration membrane. Desalination 2000;130:165.

[9] Saitua H, Campderros M, Cerutti S, Padilla AP. Effect of operating conditions in removal of arsenic from water by nanofiltration membrane. Desalination 2005;172:173–80.

[10] Nguyen VT, Vigneswaran S, Ngo HH, Shorn HK, Kandasamy J. Arsenic removal by a membrane hybrid filtration system. Desalination 2009;236:363–9.

[11] Sato Y, Kang M, Kamei T, Magra Y. Performance of nanofiltration for arsenic removal. Water Res 2002;36:3371–7.

[12] Pal P, Sikder J, Roy S, Giorno L. Process intensification in lactic acid production: a review of membrane-based processes. Chem Eng Process Process Intensif 2009;48: 1549–59.

[13] Pal P, Chakrabortty S, Roy M. Arsenic separation by a membrane-integrated hybrid treatment system: modeling, simulation and technolo-economic evaluation. Separ Sci Tech 2012;47:1091–101.

[14] Pal P, Chakrabortty S, Linnanen L. A nano-filtration-coagulation integrated system for separation and stabilization of arsenic from groundwater. Sci Total Environ 2014;476–477:601–10.

[15] Pal BN. Granular ferric hydroxide for elimination of arsenic from drinking water. In: Ahmad MF, Ali MA, Adeel Z, editors. BUET-UNU international workshop on technologies for arsenic removal from drinking water, Dhaka, Bangladesh, 5-7 May; 2001. p. 59–68.

[16] Song S, Lopez-Valdivieso A, Hernandez-Campos DJ, Peng C, Monroy-Fernandez MG, Razo-Soto I. Arsenic removal from high-arsenic water by enhanced coagulation with ferric ions and coarse calcite. Water Res 2006;40:364–72.

[17] Geucke T, Deowan SA, Hoinkis J, Patzold Ch. Performance of a small-scale RO desalinator for arsenic removal. Desalination 2009;239:198–206.

[18] Fagarassy E, Galambos I, Bekassy-Molnar E, Vatai Gy. Treatment of high arsenic content wastewater by membrane filtration. Desalination 2009;240:270–3.

[19] Strathmann H. Electrodialysis. In: Ho WS, Sirkar KK, editors. Membrane handbook. New York: Van Nostrand Reinhold; 1992. p. 217–54.

[20] Schafer AI, Nghiem DI, Waite TD. Removal of natural hormone estrone from aqueous solutions using nanofiltration and reverse osmosis. Environ Sci Technol 2003;37(1):182–8.

[21] Pal P, Ahammad Z, Pattanayek A, Bhattacharya P. Removal of arsenic from drinking water by chemical precipitation—a modelling and simulation study of the physical-chemical processes. Water Environ Res 2007;79(4):357–66.

[22] Driehaus W, Jekel M. Determination of As (III) and total inorganic arsenic by on-line pre-treatment in hydride generation atomic absorption spectrophotometry. Fresenius J Anal Chem 1992;343(4):356–62.

[23] Braghetta A. The influence of solution chemistry and operating conditions on nanofiltration of charged and uncharged organic macromolecules. Chapel Hill: University of North Carolina; 1995.

[24] Pagana AE, Sklari SD, Kikkinides ES, Zaspalis VT. Microporous ceramic membrane technology for the removal of arsenic and chromium ions from contaminated water. Micropor Mesopor Mat 2008;110:150–6.

[25] Xia S, Dong B, Zhang Q, Xu B, Gao N, Causseranda C. Study of arsenic removal by nanofiltration and its application in China. Desalination 2007;204:374–9.

[26] Hsieh L, Weng Y, Huang C, Li K. Removal of arsenic from groundwater by electro-ultrafiltration. Desalination 2008;234:402–8.

[27] Hagmeyer G, Gimbel R. Modelling the rejection of nanofiltration membranes using zeta potential measurements. Sep Sci Technol 1999;15:19–30.

[28] Labbez C, Fievet P, Szymczyk A, Vidonne A, Foissy A, Pagetti J. Retention of mineral salts by a polyamide nanofiltration membrane. Sep Purif Technol 2003;30:47–55.

[29] Bowen WR, Welfoot JS. Modeling the performance of membrane nanofiltration critical assessment and model development. Chem Eng Sci 2002;57:1121–37.

[30] Chakrabortty S, Pal P. Arsenic removal from contaminated groundwater by membrane-integrated hybrid plant: optimization and control using visual basic platform. Environ Sci Pollut Res 2014;21(5):3840–57.

[31] Bruggen BV, Vandecasteele C. Removal of pollutants from surface water and groundwater by nanofiltration: overview of possible applications in the drinking water industry. Environ Pollut 2003;122:435–45.

CHAPTER 5

Arsenic Removal by Membrane Distillation

Contents

Groundwater Arsenic Remediation
http://dx.doi.org/10.1016/B978-0-12-801281-9.00005-9

5.1 PRINCIPLES OF MEMBRANE DISTILLATION

Membrane distillation (MD) is a thermally driven membrane separation process, where the driving force is the vapor pressure difference created by temperature difference across the membrane. Separation of components by membrane distillation is based on the principle of vapor liquid equilibrium. The term membrane distillation arises from similarity of the process to conventional distillation where both processes rely on vapor–liquid equilibrium as the basis for molecular separation and both need a supply of latent heat of vaporization for the phase change from liquid to vapor.

In the MD process, water vapor transports through a microporous hydrophobic membrane, which must not be wet and should allow only the vapor and noncondensable gases to pass through its pores. The membrane material is water repellent, so liquid water cannot enter the pores unless a hydrostatic pressure, exceeding the so-called liquid entry pressure of water (LEPw), is applied. In the absence of such a pressure difference, a liquid–vapor interface is formed on the liquid contact side of the membrane pores. More specifically, it can be said that the liquid feed to be treated by MD must be in direct contact with one side of the membrane, and this liquid does not penetrate inside the dry pores because hydrophobic nature prevents liquid solution from entering into the pores. If a temperature difference is maintained through the membrane, a water vapor pressure difference appears that is the driving force for the diffusion of vapor through membrane pores. Consequently, water molecules evaporate at the hot interface, cross the membrane in vapor phase, and condense in the cold side, either in a liquid (in case of direct contact membrane distillation), on a cooling surface (in case of air gap membrane distillation), or in a condenser (either in vacuum membrane distillation or sweeping gas membrane distillation process), giving rise to a net transmembrane water flux. It is worth pointing out that the water vapor pressure difference may have a contribution due to a concentration difference. The process is operated at near-atmospheric pressure and at low temperature.

5.2 ADVANTAGES OF MEMBRANE DISTILLATION

Advantages of MD over other processes are described as follows.

5.2.1 MD over conventional distillation

The advantages of MD over conventional distillation include the following:

- MD operates at low temperatures with involvement of very low external energy. In conventional distillation the feed solution needs to be heated to its boiling point, whereas that is not necessary for MD. Low-grade heat, industrial waste heat, furnace/burner flue gas heat, or desalination waste heat and alternative energy sources such as solar and geothermal energy can be coupled with MD systems for a cost-efficient, energy-efficient liquid separation system. Low feed temperatures and pressures reduce chemical interaction between membrane and process solutions.
- Capability of effecting separation of components at low temperature makes MD a viable process for the concentration of heat-sensitive substances in the food and pharmaceutical industries.
- As the required operating feed temperatures typically range from 30 °C to 80 °C, it is not necessary to heat the process liquids above their boiling temperatures. Low operating temperature allows selection of low-cost material and nonmetallic materials like polycarbonate, which help in preventing heat loss through equipment surfaces.
- Conventional distillation relies on vapor–liquid contact with high vapor velocities apart from high temperature, whereas MD requires only hydrophobic microporous membrane to support a vapor–liquid interface, and the system can have modular design with huge operational

flexibility and compactness. As a result, the size of MD process equipment for handling same volume of liquid feed is much smaller than that required in conventional distillation. All these result in lower capital as well as operating costs and reduced space requirements.

5.2.3 MD over pressure-driven membrane processes

MD generally employs microporous hydrophobic membrane of the pore sizes in the range of microfiltration (0.1 to 1 μm). Major advantages of MD over pressure-driven processes such as reverse osmosis (RO), nanofiltration (NF), ultrafiltration (UF), and microfiltration (MF) can be focused in the following points.

5.2.3.1 Low operating temperature and hydrostatic pressure

The MD process can be carried out under mild operation conditions. Feed solutions having temperatures much lower than its boiling point under pressure near atmosphere can be used. Operating pressures in MD are generally on the order of zero to a few hundred kPa, relatively low compared to pressure-driven processes. Lower operating pressures in MD means lower equipment costs, increased process safety, and reduced mechanical demands (i.e., resistance to compaction) on microporous membranes. In addition, one important benefit from low operation pressure is the reduction of membrane fouling.

5.2.3.2 Solute rejection

Another benefit of MD stems from its efficiency in terms of solute rejection. Since MD operates on the principles of vapor–liquid equilibrium, 100% (theoretical) of ions, macromolecules, colloids, cells, and other nonvolatile constituents are rejected; whereas pressure-driven processes such as RO, UF, pervaporation (PV), and MF have not been shown to achieve such high levels of rejection.

5.2.3.3 Membrane selectivity

In MD, the membrane acts merely as a support for vapor-liquid interfaces at the entrance of the pores and does not distinguish between components on a chemical basis in the feed solution, nor does it act as a screen. Therefore, the role of the membrane in MD is minimum when we compare it with RO, UF, and MF. Due to this reason, most MD membranes can be prepared from chemically resistant polymers such as polytetrafluoroethylene (PTFE), polypropylene (PP), and polyvinylidene difluoride (PVDF).

5.2.3.4 Membrane fouling

In general, membrane fouling occurs by the deposition and accumulation of undesirable materials (i.e., organic compound, inorganic compound, or a combination of both) on membrane surfaces. In MD, the membrane surface at the feed side is in direct contact with the vapor of the vapor–liquid interface. Therefore, the possibility of membrane fouling is much less compared to other membrane separations. In MD, membrane pore clogging is also less because the pores are relatively large compared to the pores of diffusional pathways in RO, NF, or UF. Membrane fouling is a serious problem in all pressure-driven membrane separation processes and will reduce flux and life span of the membrane. The feed water does not require extensive pretreatment to prevent membrane fouling as seen in pressure-based membrane processes. The MD process is the least affected membrane process by membrane fouling. Concentration polarization, which has a major effect on other pressure-driven membrane processes, has a negligible effect on the MD process. Other advantages are low sensitivity of such membranes to pH and concentration of salts and availability of membranes having excellent mechanical and chemical resistance properties.

5.2.4 Membrane distillation over osmotic membrane distillation (OMD)

In the 1980s, it was observed [1] that the same MD membranes could be used in a different isothermal process termed osmotic membrane distillation (OMD). OMD is a new membrane process, which is very similar to the MD process. The similarity is that both use the same membrane of hydrophobic nature as the supporters of the liquid-vapor interface for the evaporation of volatile components, and same mass transfer mechanisms for causing the mass transfer. The only difference between them is the way by which the driving force for mass transfer is exerted. The OMD process, which has been developed dynamically in recent years, can also be included in this group. In OMD, the feeding solution temperature is low and close to the temperature of the solution flowing on the other side of the membrane. The driving force for the transmembrane flux is associated with transmembrane osmotic pressure (water activity) difference influenced by the two aqueous solutions of different concentrations at both sides of the membrane. The OMD is similar to direct contact MD (DCMD). Therefore, the equations derived for the DCMD variant are successfully applied also for the description of the vapor transport in the OMD process [2]. However, several differences between DCMD and OMD variants exist. The main difference is that the driving force is formed by the temperature gradient in MD, whereas in OMD, this is by concentration gradient. The major disadvantage of the OMD process is low fluxes. OMD can be performed at room temperature, so it can apply for treating heat sensitive substances, such as fruit juice, jam, enzyme, protein, and so on. An OMD pilot plant in Australia has been set up for the concentration of fruit juice and vegetable juice. The capacity of this plant is 100 L/hr and the concentration of 65–70% is achieved [3].

5.3 LIMITATIONS OF MEMBRANE DISTILLATION

The MD process has some limitations also that arise due to wetting of the membrane or the presence of multiple volatile components. The hydrophobic pores of the membrane should not be wetted by the solutions below a certain level of pressure, called the liquid entry pressure (LEP). Presence of organic solutes in the aqueous solutions decreases liquid entry pressure. With the increase of concentration of the organic solutes spontaneous wetting of the membrane occurs. Therefore, MD can be applied to only a narrow range of concentrations of organic solutes. This range depends also on the membrane and the temperature of the solutions. The process solutions must be aqueous and sufficiently dilute to overcome this problem of pore wetting.

Normally, 100% theoretical rejection of nonvolatile solutes in feed aqueous solution is achieved in MD process since only water vapor passes through the membrane. But such a high degree of separation is not possible when two or more components vaporize and permeate through the membrane, and separation efficiency will depend on relative volatility of the components.

5.4 MEMBRANE MATERIALS

Large varieties of membranes, such as polytetrafluoroethylene (PTFE), polypropylene (PP), and polyvinylidene fluoride (PVDF), that match the requirements of MD are commercially available for application in different modules like flat-sheet, plate-and-frame, spiral-wound, and capillary or tubular.

5.4.1 Polyvinylidene fluoride

The repeat unit of PVDF polymer is $-(CF_2-CH_2)-$. PVDF is of special interest in membrane distillation processes because of its high melting point and good resistance to abrasion, temperature, oxidation, gamma radiation, and solvents. PVDF is resistant to most inorganic and organic acids and can be used in a wide pH range. It is also stable in aromatic hydrocarbons, alcohols, tetrahydrofurane, and halogenated solvents. PVDF is a semicrystalline with a very low T_g (glass transition temperature) (−40 °C), which makes it quite flexible and suitable for membrane application in temperatures ranging from −50 to 140 °C, just prior to its melting temperature. Although stable in most organic solvents, PVDF is soluble in dimethyl formamide (DMF), dimethyl acetamide (DMAc), N-methyl pyrrolidone (NMP), and dimethyl sulfoxide, permitting membrane preparation by a phase inversion process. In early patent [4] on PVDF membranes, solution containing about 20% PVDF in DMAc were cast and immersed in a methanol bath. Preparation of PVDF membranes can be done by dissolving the polymer in boiling acetone and immersing it in a cold water/acetone bath [5]. Synthesis of MD membrane may also be done by dissolving the polymer PVDF in DMF (8–15%) or DMAc (8%) in the presence of LiCl at elevated temperature, and then the solutions may be cast on a glass plate, followed by immersion of the plate within the film into a water bath at 277 K [6].

The membranes from PVDF are prepared by the thermal phase separation process [7] or wet phase inversion process.

5.4.2 Polytetrafluoroethylene

The monomer unit in PTFE is $-(CF_2-CF_2)-$. The PTFE membranes are formed by the stretching and heating process [6]. Stretching is part of the preparation process of commercial membrane Gore–Tex PTFE membranes. Cold drawing has also been used [8] for membrane preparation starting from crystalline polymers. Another preparation method is solvent stretching [9], where the precursor film is brought in contact with a swelling agent and stretched. The swelling agent is removed while the film is maintained stretched to render the film microporous. Other processes use sequential "cold" and "hot" stretching steps [10].

Commercial membrane prepared by stretching is Gore–Tex. The polymer here is PTFE, which is what makes the membrane extremely inert and thus convenient for processing even harsh streams. Processing PTFE is possible only by paste extrusion. In paste forming, the polymer is mixed with a lubricant such as odorless mineral spirits naphtha or kerosene. The lubricant component is removed by heating to 327 °C. Above this temperature, sintering would lead to a dense PTFE film. After lubricant removal, the PTFE film is submitted to uniaxial or biaxial stretching, giving rise to an interconnected pore structure. The process was proposed by Gore [11] and the resulting porous film is today a successful product in the membrane and textile industry. For uniaxial stretching, the unsintered film from the paste extrusion is fed to a machine with heating roles, where one role is driven faster than the previous one to input stress and induce the pore formation. The difference in speed determines the amount of stretch. Additionally, in the Gore patent a biaxial stretching is performed using a pantograph. A special characteristic of the Gore membrane is that, since PTFE is very hydrophobic, (liquid) water must not be allowed to wet the membrane and its transport is hindered. On the other hand, water vapor can freely pass through the micropores, making the film suitable for transporting impermeable cloths. However due to their inertness, PTFE membranes are also interesting during the processing of aggressive streams.

5.4.3 Polypropylene

The repeat unit of PP polymer is $-(CH_2-CH(CH_3))-$. Stretching is also part of the preparation process of the commercial membrane Celgard. The Celgard membrane is made of PP, which is a low-cost and quite inert polymer. It is resistant to extreme pH conditions and is insoluble in most solvents at room temperature. It swells in polar solvents such as carbon tetrachloride. No solvent is required for the preparation of the membrane. It involves the extrusion of PP films with high melt stress to align the polymer chains and induce the formation of lamellar micro crystallites when cooling. The film is then stretched by 50–300%, just below the melting temperature. Under stress, the amorphous phase between the crystallites deforms, giving rise to the slit-like pores of the Celgard membrane. The film is then cooled under tension. PP is highly hydrophobic and several surface treatments have been proposed to improve the hydrophilicity of PP membranes. Incorporation of surfactants is used to make Celgard membranes more hydrophilic. Polypropylene exhibits the smallest contact angle among the polymers used for preparation of MD membranes.

5.4.4 Membrane characteristics

From the principles of membrane distillation, it is obvious that the membrane material must be porous and hydrophobic, have high permeability (flux), and good thermal stability. It should also have low resistance to mass transfer, high liquid entry pressure of water to maintain the membrane pores dry, low thermal conductivity to prevent heat loss through membrane matrix, and excellent chemical resistance to feed solutions. The relationship between the molar flux through the membrane pore and the different membrane characteristic parameters is given by the following equation [12]:

$$N \propto \frac{r^a \varepsilon}{\tau \delta} \tag{5.1}$$

where N is the molar flux, r is the average pore size of the membrane, and a is a factor whose value equals 1 or 2 for Knudsen diffusion and viscous fluxes, respectively. δ is the membrane thickness, ε is the membrane porosity, and τ is the membrane tortuosity. Therefore, a microporous membrane for MD is generally characterized by four parameters: r, δ, ε, and τ. Table 5.1 shows the major characteristics of the membranes for use in membrane distillation [13,14].

Membrane characteristics, such as porosity, distribution of pore size, thickness, tortuosity, and mean pore diameter play important roles in membrane distillation performance. The gas permeation (GP) experiment is one of the well-known experimental methods used to determine the characteristics of such porous membranes.

5.4.4.1 Membrane pore size

Membranes with pore sizes ranging from 0.1 to 1 μm can be generally used in MD. According to Eq. (5.1), membrane flux through a pore of average pore size r is proportional to r^a when the membrane porosity (ε), the membrane tortuosity (τ), and the membrane thickness (δ) remain constant. With an increase in pore size, the transmembrane flux enhances, and also the chance of membrane wetting increases. Therefore, an optimum value of pore size is required to be determined for each MD application. This optimum value is determined based on the kinetic theory of gases [14]. The optimum membrane pore size depends on the type of feed solution to be treated. According to the Dusty Gas Model, the transport of vapors or gases through porous membrane can be described by several mechanisms [15]. The selection of the most suitable mechanism strongly depends on the comparison of mean free path (λ) and pore diameter (d_p).

Table 5.1 Typical Values of the Major Characteristics of Membranes Used in MD

Mean pore diameter (mm)	Porosity (%)	Thickness (mm)	Tortuosity
0.1–1.0	30–90	20–200	1.5–2.5

5.4.4.2 Membrane porosity and pore size distribution

The membrane porosity is defined as the ratio of the volume of voids in the membrane to the total volume (voids plus solid). The porosity of a membrane mainly depends on the technology of membrane preparation. Membrane porosity up to 90% is commercially available. Membrane porosity is considered one of the most important membrane parameters because higher membrane porosity provides higher permeate flux due to greater surface area available for evaporation. Membrane having high porosity reduces the conductive heat loss, which can be explained by the equation,

$$h_m = \varepsilon h_{mg} + (1 - \varepsilon) h_{ms} \qquad (5.2)$$

where h_m, ε, and h_{mg} and h_{ms} are the membrane heat transfer coefficient, the membrane porosity, and the heat transfer coefficients of the gas (vapor) within the membrane pores and the solid membrane material, respectively. The value of the conductive heat transfer coefficient (h_{mg}) of the gases (vapor) is smaller than that (h_{ms}) of the hydrophobic polymer used for membrane preparation. Therefore, h_m can be minimized by maximizing the membrane porosity. In general, membrane porosity in MD lies between 30% and 85% [14].

Various pore shapes rather than uniform pore size are observed in the membranes employed in MD. Phattaranawik et al. [41] observed various pore shapes (such as elliptical, circular, and slit) of the membranes made of PVDF and PTFE from the images. From the mass transfer point of view and the wetting phenomenon in the MD process, a sharp pore size distribution is set between 0.3 and 0.5 μm [16]. The pore size distributions of the membranes are determined by field emission scanning electron microscopy (FESEM) and the image analysis program. As the pore size in MD membranes is not uniform, more than one mechanism of mass transport can simultaneously occur to a different extent. Most literatures have employed the average pore size in the gas transport model to calculate the fluxes. Gaussian (symmetric) and logarithmic (asymmetric) distributions have also been applied for evaluation of the effect of shape of pore size distribution on permeate flux [17].

From the Laplace (Cantor) Eq. (5.3), it is clear that the membrane with larger pore size has relatively lower liquid entry pressure (LEP), which leads to easy wetting during operation. In order to prevent the wetting phenomenon in a DCMD process, the membrane pore size should be smaller than 0.5 μm [18]. From this the upper limit for the average pore diameter in MD membranes should be 0.5 μm. For a given pore size, a critical penetration pressure P_c exists. Liquid phase is transported across the membranes for applied hydrostatic pressure higher than the P_c value.

5.4.4.3 Membrane thickness and pore tortuosity

When average pore size (r) and the membrane porosity (ε) remain constant, membrane flux through a pore is inversely proportional to the transport path length through the membrane, $\tau\delta$ as predicted by Eq. (5.1). From Eq. (5.1), it is clear that permeate flux in MD is inversely proportional to membrane thickness (δ), as observed in other membrane separation processes, and this is valid in the case of all MD configurations except air gap membrane distillation (AGMD), where the stagnant air gap offers more resistance to mass transfer than what the thick membrane offers [19]. On the other hand, heat loss through membrane matrix by conduction is also inversely proportional to membrane thickness (δ). Therefore, the membrane should be as thick as possible to get better heat efficiency. A thin membrane induces the problem of negative flux (mass transfer takes place from permeate side to feed side) for aqueous solutions with significant osmotic pressure [20]. Thus to ensure reasonable flux and low heat loss, there exists an optimum membrane thickness that needs to be determined. Through computer simulation an optimum value is found to be within the range of 30–60 μm [17].

Membrane tortuosity is defined as the ratio of pore length to membrane thickness. Generally, vapor molecules in the MD process move through the tortuous channels in the membrane instead of straight channels across the membrane. The actual channels are irregular in shape, have a variable cross section and orientation, and are highly interconnected. However, to calculate the average parameters (e.g., velocity of vapor molecules), it is assumed that the actual channels may be effectively replaced by a set of identical parallel conduits of constant cross section. Because of the difficulties in measuring its real value for any of the microporous membrane, pore tortuosity is used in MD process as a correction factor for the prediction of MD flux. However, the pore tortuosity is determined by the gas permeation test [21].

5.4.4.4 Liquid entry pressure (LEP) and membrane wetting

LEP is the minimum hydrostatic pressure that is required on the solution side to overcome the hydrophobic forces of the membrane for penetration of the liquid into the membrane pores. Membranes for use in membrane distillation application are very much characterized by LEP, where the desired property of the membrane is to prevent wetting of the pores. The higher the LEP of a membrane the better it is in preventing membrane wetting.

The wettability of these membranes depends on several factors such as the liquid surface tension, the liquid–membrane contact angle (higher than 90°), and the size and shape of the pores. The rough idea of a membrane wetting can be realized by using the Laplace (Cantor) equation, which provides the relationship between the membrane's largest allowable pore size (r_{max}) and operating conditions [12].

$$\text{LEP} > \Delta P_{\text{interface}} = P_{\text{liquid}} - P_{\text{vapor}} = -\frac{2B\gamma_l\cos\theta}{r_{max}} \tag{5.3}$$

where B is a geometric factor determined by pore structure (for instance, $B=1$ for cylindrical pore), γ_l is the liquid surface tension, and θ is the liquid/solid contact angle. When $\Delta P_{\text{interface}}$ is greater than LEP, the liquid can enter the membrane pores. This membrane pore penetration is known as wetting of a membrane. The vapor–liquid equilibrium at the membrane interface is lost after wetting of it.

5.4.4.5 Liquid solid contact angle and liquid surface tension

To prevent wetting of the membrane pores, LEP should be high. High LEP is ensured by high hydrophobicity (i.e., large water–solid contact angle) and low surface energy. Surface tension of the liquid in direct contact with the membrane should be large in MD membrane. In general, the liquid–solid contact angle must be greater than 90° for the system to be used in MD. Contamination of feed solution with strong surfactants reduces the value of $\gamma_l\cos\theta$, which leads to the decrease of LEP and thus permits the feed to penetrate through the membrane pores. Therefore, care must be taken to prevent equipment and solution from being contaminated by detergents and other surfactants.

Figure 5.1 illustrates the cross-sectional view of supported straight cylindrical pores in contact with an aqueous solution to show the contact angle (θ) and how the vapor–liquid interfaces are supported at the pore openings.

5.4.4.6 Fouling and scaling

The other main reasons for MD membrane wetting are fouling and scaling phenomena. A deposit formed on the hydrophobic surface of membrane causes wetting of the pores adjacent to the deposit [22].

5.4.4.7 Permeate quality and membrane wetting

According to the quality of filtrate, we can distinguish four degrees of membrane wetting: nonwetted, surface-wetted, partial-wetted, and wetted (shown in Figure 5.2). In surface-wetted membrane (Figure 5.2B), the

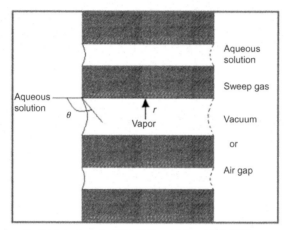

Figure 5.1 Vapor liquid interface in MD. *(Adapted from Ref. [12]).*

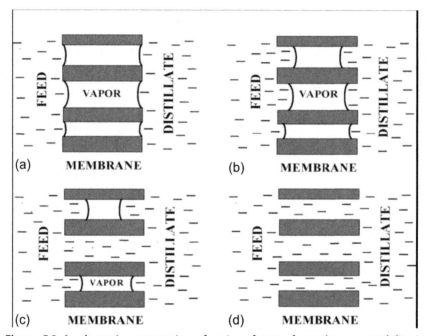

Figure 5.2 A schematic presentation of various forms of membrane wettability in MD process: (A) nonwetted; (B) surface-wetted; (C) partial-wetted; and (D) wetted.

liquid enters the pores up to a significant depth, maintaining a gaseous gap between the feed and within the membrane. As a result, a high purity filtrate in surface-wetted membrane is produced as observed in nonwetted membrane. In partial-wetted membrane (Figure 5.2C), a deterioration of the filtrate quality occurs as some pores are wetted. In this case, a slightly higher hydraulic pressure on the permeate side is maintained to prevent a leakage of the feed, and in the case when a wetted membrane area is not too large, the MD process may be still continued [23].

Membrane distillation (MD) operations generally have been carried out with porous hydrophobic membranes, initially manufactured for microfiltration purposes. To avoid pore wetting, the porous membrane must be hydrophobic with high water contact angle and small maximum pore size.

There have been some available microporous hydrophobic membranes made by PTFE, PP, and PVDF that meet these requirements. Table 5.2 shows some of the commercial membranes commonly used in MD processes together with some of their characteristics.

Moreover, these polymers exhibit excellent chemical resistance and good physical and thermal stability. Thermal conductivity of membrane materials (PVDF, PTFE, and PP), gas (air), and vapor (water) involved in MD is shown in Table 5.3 [13].

As in most cases in MD, the pores in the hydrophobic membrane are filled with air and water vapor, and their thermal conductivities, k_{water} and k_{air}, depend on temperature (T) obeying the following equations at atmospheric pressure:

$$k_{water} = 2.72 \times 10^{-3} + 5.71 \times 10^{-5}\, T \tag{5.4}$$

$$k_{air} = 2.72 \times 10^{-3} + 7.77 \times 10^{-5}\, T \tag{5.5}$$

The thermal conductivities of hydrophobic membrane polymers (PVDF, PTFE, and PP) are found to be large ranges of span due to thermal conductivities of polymers, which depend upon both temperature and degree of crystallinity. From the preceding tables and equations, it is clear that heat lost by conduction can be reduced by increasing the porosity (ε) of the membrane.

The conductive heat loss through the membrane can be reduced by using membrane material with low thermal conductivity, membrane with high porosity, thicker membrane, or composite porous hydrophobic/hydrophilic membrane. A composite porous hydrophobic/hydrophilic membrane contains a very thin hydrophobic layer responsible for mass transfer and a thick hydrophilic layer, in which the pores are filled with water that prevent heat loss through the overall membrane. A composite porous hydrophobic/hydrophilic membrane not only provides low thermal conductivity but also has a high permeability.

Table 5.2 Commercial Membranes Commonly Used in Membrane Distillation

Manufacturer	Material	Description	Membrane trade name	Thickness (without support) (μm)	Average pore size (μm)	Porosity (without support) (%)
Millipore	PVDF	Flat-sheet membranes	Durapore	110	0.45	75
			Durapore	140	0.22	75
Gelman Inst. Co.	PTFE	Flat-sheet membranes supported on polymer fabric	TF200	60	0.20	60
			TF450	60	0.45	60
			TF1000	60	1.00	60
Gore	PTFE	–	Gore–Tex	Wide range	Wide range	Wide range
Akzo Nobel Microdyn	PP	Capillary membrane	S6/2 MD020CP2N	45.0	0.2	70
Hoechst – Celanese	PP	–	Celgard 2400	25	0.02	38
	PP	Tube	Celgard X–20	25	0.03	35

Table 5.3 Thermal Conductivity of Membrane Materials, Gas, and Water Vapor Involved in MD

Temperature (K)	PVDF (W/(m K))	PTFE (W/(m K))	PP (W/(m K))	Air (W/(m K))	Water vapor (W/(m K))
296	0.17–0.19	0.25–0.27	0.11–0.16	0.026	0.020
348	0.21	0.29	0.20	0.030	0.022

5.5 MEMBRANE DISTILLATION CONFIGURATIONS

In the MD process, a heated aqueous feed solution is brought in contact with one side (feed side) of a hydrophobic, microporous membrane. After the evaporation of volatile molecules, to be separated from the feed, at the hot feed side, transport of vapor through dry pores of hydrophobic membranes occurs due to a vapor pressure difference across the membrane, which is the driving force (also known as transmembrane vapor pressure difference) to drive flux. Depending on the method by which the permeating vapor is recovered from the pores at the other side of the membrane (permeate side), the MD processes may be carried out with different configurations.

5.5.1 Direct contact membrane distillation (DCMD)

In DCMD configuration, direct contact means that both the feed and the permeate liquid (flowing across the two chambers, respectively) are in direct contact with the membrane in the chambers. The DCMD configuration is shown in Figure 5.3.

In this configuration, volatile molecules evaporate at the liquid/vapor interface in the feed side, passes the membrane pores, and then condenses inside the membrane module in the cold permeate side. DCMD configuration is applied for higher flux in spite of large portion heat losses due to conduction through the membrane matrix, which is higher than other MD configurations. Another restriction occurs in DCMD, and the cooling liquid should be highly purified because cooling liquid is directly in contact with the permeate side of the membrane surface. The vapor pressure-driven transport of vapor in DCMD results from the difference in the temperature and composition of solutions in the layers adjoining the membrane.

5.5.2 Air gap membrane distillation (AGMD)

In the AGMD configuration (shown in Figure 5.4), a condensing surface is separated from the membrane by a stagnant air gap in the permeate side.

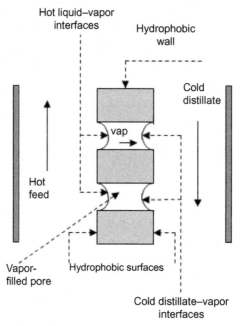

Figure 5.3 DCMD configuration.

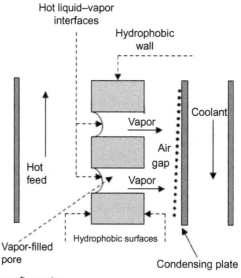

Figure 5.4 AGMD configuration.

In this case, the evaporated volatile molecules first cross the membrane pores, then cross the air gap, and finally condense over a cold surface inside the membrane module.

Air gap width in AGMD configuration plays a major role in AGMD flux and conductive heat loss through membrane. The flux in AGMD decreases with the increase of air gap width, and conductive heat loss rapidly decreases as the air gap width increases, but then gradually slowly decreases with further increases of air gap width and ultimately remains constant [24]. This phenomenon is attributed to the differences in the acid–air and water–air diffusion rates in the air gap.

5.5.3 Sweeping gas membrane distillation (SGMD)

After crossing the membrane pores from the feed side, permeating vapor is carried by a cold inert sweeping gas in the permeate side of the membrane and condensation takes place outside the membrane module in a condenser. This MD configuration (shown in Figure 5.5) is known as sweeping gas membrane distillation (SGMD).

This configuration was introduced to provide an intermediate solution between the DCMD and the AGMD configurations [25]. The SGMD configuration combines a relatively low conductive heat loss with a reduced mass transfer resistance. The SGMD flux is independent of the temperature

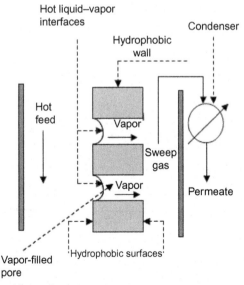

Figure 5.5 SGMD configuration.

of the sweep gas. However, the flux in SGMD increases through a maximum as the gas velocity increases, but then begins to decrease [26].

5.5.4 Vacuum membrane distillation (VMD)

In this configuration, permeating vapor condenses outside the membrane module after transportation of the vapor under vacuum applied in the permeate side of the membrane module by means of a vacuum pump as illustrated in Figure 5.6. The applied vacuum pressure is lower than the saturation pressure of volatile molecules to be separated from the feed solution [27].

5.5.5 VMD versus pervaporation

Although membrane distillation and pervaporation share some common characteristics, such as phase change and external permeate condensation, significant differences exist between them. A microporous membrane is used in VMD when acting as a barrier to hold the liquid/vapor interfaces at the pore entrance; the separation is determined by vapor liquid equilibrium and not necessary to be selective as required in pervaporation. On the other hand, pervaporation employs a dense membrane, and separation is based on the relative solubility and diffusivities of each component in the material; that is, separation is determined by selective sorption and diffusion through the membrane, and subsequent evaporation at the downside of the membrane.

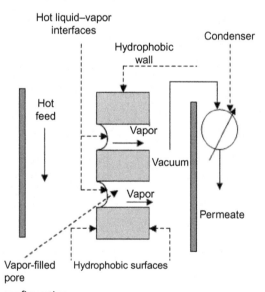

Figure 5.6 VMD configuration.

Another difference is that VMD fluxes are much higher than pervaporation fluxes. Although the largest degree of the separation occurred by the vapor–liquid equilibrium conditions at the membrane–solution interface, the VMD membrane may play a role in some selectivity based on individual Knudsen diffusivities of diffusing species, depending on the membrane pore size and the system operating conditions.

5.5.6 Comparison between different MD configurations

Figure 5.7 shows a comparison between the numbers of gathered papers (up to 2005) published in journals for each MD configuration together with the corresponding number of papers involving theoretical models.

In spite of higher heat lost by conduction in DCMD configuration compared to other configurations, more than 60% of the total published papers (Figure 5.7) are focused on DCMD configuration since it eliminates the need for a separate condenser like in SGMD and VMD configurations. Condensation takes place inside the membrane module, and the process is simplest to operate. Around 16% of total published papers deal with AGMD configuration due to the fact that imposed air gap incurs lower permeate flux compared to other MD configurations, though introduced air gap in AGMD considerably reduces heat loss by conduction and temperature polarization, which are higher in the DCMD configuration. An examination of the literature shows that the DCMD has been the most frequently considered. On the contrary, very little attention has been paid to the application in the field of SGMD because a gas source is required, the costs associated with the transported gas are very high, the permeating vapor must be collected in an external condenser, and the condenser implies a lot of work,

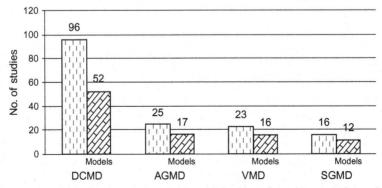

Figure 5.7 Number of papers (up to 2005) published in refereed journals for each MD configuration and corresponding papers involving theoretical models [14].

because a tiny volume of permeate is vaporized in a large volume of sweep gas. In spite of the fact that VMD shows higher permeate flux and negligible heat transfer by conduction through the membrane, very little work has been done in the field of VMD (dealing with only 14% of the relevant references) together with SGMD. This is probably due to the fact that the risk of membrane pore wetting is high. Special care must be taken in VMD to prevent membrane wetting, because $\Delta P_{interface}$ ($P_{liquid} - P_{vapor}$) is typically higher in VMD than in the other MD configurations [12]. For this reason, in VMD, the transmembrane hydrostatic pressure must be kept below the minimum feed liquid entry pressure (LEP) of the membrane. Temperature polarization in DCMD, AGMD, and SGMD show on both sides of the membranes, instead of only the feed side as shown in the VMD configuration.

5.6 BASIC DESIGN CRITERIA OF MEMBRANE MODULES AND CHANGES OVER TIME

Plate-and-frame modules, spiral-wound modules, and tubular membrane modules are most widely used. The basic requirements of MD modules are high feed and permeate flow rates with high turbulence, high rates of heat and mass transfer between the bulk liquid and the liquid-membrane interface, low concentration polarization to prevent membrane wetting and scaling, low pressure drop, high packing density (area-to-volume ratios), large membrane surface area, having good heat recovery function, and thermal stability. A membrane module should be well designed in such a way that these requirements are fulfilled. MD modules are designed by using flat-sheet membranes, capillary membranes, and hollow-fiber membranes. These membranes are fabricated from different polymers including PP, PTFE, or PVDF. Generally in different MD studies, plate-and-frame modules or spiral-wound modules use flat-sheet membranes, whereas tubular modules use capillary membranes. The nonavailability of the industrial MD modules is one of the limitations of industrial implementation of the MD process on a large scale.

Most of the laboratory scale MD modules are fabricated from flat-sheet membranes because they are easily removed from their modules for cleaning, examination, or replacement. Different types of membranes fabricated either from PP, PTFE, or PVDF can be used for testing in the same membrane module. Tubular (capillary) membranes, on the other hand, cannot easily be removed from the membrane modules, as they are a permanent integral part of the module. Though flat-sheet membranes are more versatile than capillary membranes, tubular membranes, from a commercial viewpoint, are a more advantageous position than flat-sheet membranes. Tubular membrane modules are more productive, with a much higher membrane surface area to module volume ratio. The shell-and-tube membrane module manufactured by Enka AG (Akzo) was one of the first commercially available modules. Within the first few years of its development in the 1960s, interest in MD had faded out largely because of poor performance in terms of flux and nonavailability of membranes and modules toward economically viable system performance. Very recently membrane distillation has regained its importance largely because of

- Availability of novel membranes with high porosities (>80%) and low thickness(<50 μm)
- Emergence of new modules manufactured by new techniques
- Possibility of using low-grade solar and geothermal energy due to requirement of low feed temperature (<100 °C)
- Better understanding of temperature and concentration polarization phenomena

Interest in MD boosted up further with the possibility of enhanced funding for research on environment-related problems and utilization of renewable energy, where the MD process has the potential of yielding environment-friendly results and solutions to environmental problems. However, low flux still stood in the way of application of membrane distillation (MD). In the very recent past, attempts have been made to purify arsenic-contaminated water using membrane distillation. In laboratory-scale direct-contact membrane distillation (DCMD) units arsenic removal has been attempted [28] using self-made PVDF capillary membranes. But a maximum permeate flux of only 20.90 kg/(m^2 hr) has been obtained. A hollow-fiber membrane module containing membrane made of polypropylene has also been used [29], again yielding low flux. An air gap membrane distillation (AGMD) module has also been investigated for removal of arsenic using a small-scale commercial prototype MD module [30].

Pal and Manna developed a new solar-driven flash vaporization membrane module and applied the same in removal of arsenic from contaminated groundwater [31]. This new design for the first time substantially enhanced flux to the level of around 50 LMH, thus addressing the issue of low flux considered as the major hurdle in application of MD. This design of the flat-sheet cross-flow solar-driven membrane module in the DCMD configuration successfully produced almost 100% arsenic-free water from contaminated groundwater in a largely fouling-free operation while permitting high fluxes under reduced temperature polarization. The encouraging results showed that the design could be effectively exploited in the vast arsenic-affected rural areas of South East Asian countries blessed with abundant sunlight particularly during the critical dry season.

5.7 HEAT TRANSFER IN MEMBRANE DISTILLATION

The mechanism of physical transport phenomena in MD involves simultaneously both heat and mass transfer. The heat transfer in MD is very important and is believed to be the rate controlling step in the MD process.

5.7.1 Temperature polarization effect

According to the classical theory of heat transfer, a thermal boundary layer develops when a fluid is in direct contact with the solid surface, as long as the temperatures of the solid surface and the fluid are different. In the MD process, feed and permeate with different temperatures are separated by a microporous hydrophobic membrane of thickness δ and therefore, a feed boundary layer of thickness δ_{ft} in the feed side and a permeate boundary layer of thickness δ_{pt} on the other side (permeate side) of the membrane are formed, as shown in Figure 5.8. Each thermal boundary layer imposes a resistance to heat transfer and also effects negatively the driving force for mass transfer. This effect leads to a decrease in the MD flux. This phenomenon is called the temperature polarization effect. Up to 80% reduction

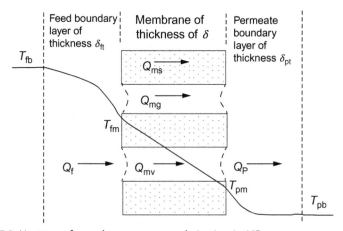

Figure 5.8 Heat transfer and temperature polarization in MD.

of driving force may be observed due to the temperature polarization effect [12]. Due to the temperature polarization effect, the feed temperature decreases from the value of T_{fb} at the bulk feed to T_{fm} at the feed/membrane surface, while the permeate temperature increases from the value of T_{pb} at the bulk permeate to T_{pm} at the permeate/membrane surface.

The temperature polarization coefficient (TPC) has been used to characterize the detrimental effect of the temperature polarization phenomenon on flux [32,33]. TPC is defined as the ratio of the transmembrane temperature to the bulk temperature difference.

$$TPC = \frac{T_{fm} - T_{pm}}{T_{fb} - T_{pb}} \tag{5.6}$$

TPC is also used to represent the loss of thermal driving force due to thermal boundary layer resistances. The value of TPC approaches unity for DCMD systems that have good arrangement of fluid dynamic, and the process is controlled by mass transfer through the membrane. TPC approaches zero for the poorly designed system that is limited by the heat transfer through the boundary layer. In well-designed systems, TPC is available in a range of 0.4–0.7 [34]. TPC is observed between 0.4 and 0.53 at low cross-flow velocity (0.23 m/sec, laminar region) and 0.87–0.92 at high cross-flow velocity (1.85 m/sec, turbulent region) [35]. High temperature results in high flux, which in turn means the requirement of a larger amount of heat to vaporize water at the membrane surface. This enhances the effect of temperature polarization due to an increased difference of temperature between the bulk feed stream and the membrane surface. Temperature polarization effects can be reduced by improving the design of flow passage, membrane arrangement, or applying turbulence promoters like mesh spacers. Temperature polarization effects are highly reduced economically under better mixing conditions to avoid pore wetting. The factors influencing TPC are the properties of the fluids, flow velocity, module geometrical shape, among others.

5.7.2 Heat transfer steps in the MD process

Membrane distillation is a complicated physical process involving both heat transfer and mass transfer. Heat transfer in MD can be considered a three-step process based on three regions (feed boundary layer, membrane, and permeate boundary layer) as illustrated in Figure 5.8: convective heat transfer in the feed boundary layer, heat transfer across the membrane, and convective heat transfer in the permeate boundary layer. Heat transfer due to mass transfer across the feed thermal boundary layer is around 4% of the total heat flux. Heat transfer due to mass transfer across the filtrate thermal boundary layer is only about 1.4% of the total heat flux [36]. Therefore, the effects of mass transfer on the heat transfers (Dufour effect) in the feed and permeate thermal boundary layers are negligible in the heat transfer steps.

5.7.3 Convective heat transfer in the feed boundary layer

The convective heat flux, Q_f, depends on the film heat transfer coefficient in the feed boundary layer, h_f, and the temperature difference between the feed bulk and the feed/membrane surface. It can be expressed mathematically as

$$Q_f = h_f(T_{fb} - T_{fm}) \tag{5.7}$$

5.7.4 Heat transfer across the membrane

Heat transfer involved across the membrane can be divided into three substeps: (1) conduction heat transfer through the membrane matrix, Q_{ms}; (2) conduction heat transfer through the gases (or air vapor) in the pores, Q_{mg}; and (3) heat transferred because of vapor migration through the membrane pores, Q_{mv}.

These heat transfer mechanisms can be expressed mathematically as follows:

Conduction through the membrane matrix:

$$Q_{ms} = \frac{k_s(1-\varepsilon)}{\delta}(T_{fm} - T_{pm}) = h_{ms}(T_{fm} - T_{pm}) \tag{5.8}$$

where k_s, δ, and ε are the thermal conductivity, thickness, and porosity of the hydrophobic membrane polymer, respectively. h_{ms} is the heat transfer coefficient related to the conductive heat flux through the membrane matrix.

Conduction through the gases entrapped in the pores:

$$Q_{mg} = \frac{k_g\varepsilon}{\delta}(T_{fm} - T_{pm}) = h_{mg}(T_{fm} - T_{pm}) \tag{5.9}$$

where k_g is the thermal conductivity of the gases trapped inside the membrane pores. h_{mg} is the heat transfer coefficient related to the conductive heat flux through the gases entrapped in the membrane pores.

Combination of both conductive heat transfers through both the membrane matrix and gases entrapped in the pores in the membrane can be evaluated from:

$$Q_{mc} = Q_{ms} + Q_{mg} = \frac{k_m(T_{fm} - T_{pm})}{\delta} = h_m(T_{fm} - T_{pm}) \tag{5.10}$$

where k_m is the thermal conductivity of polymeric membrane, which can be determined on the basis of the membrane material data:

$$k_m = (1 - \varepsilon)k_s + k_g\varepsilon \tag{5.11}$$

and h_m is the conductive heat transfer coefficient of the membrane, which can be determined by:

$$h_m = h_{ms} + h_{mg} = \frac{k_m}{\delta} \tag{5.12}$$

Equations (5.9) and (5.10) imply that the heat transfer coefficient of the membrane depends on its thickness and porosity, the materials of which the membrane is fabricated, the heat conductivity of the membrane material, and the air or vapor trapped within the membrane pores. From Eq. (5.11), it is clear that the value of thermal conductivity of water vapor or air trapped within the membrane pores is smaller than that of the membrane material; therefore, the value of h_m can be reduced by increasing the membrane porosity, ε, which reduces heat lost by conduction through the membrane. The internal heat lost by conduction, Q_{mc} is negligible in the VMD configuration due to the applied vacuum at the permeate side when it is compared with other MD configurations. Convective heat transfer through the membrane pores is negligible and its value is 0.6% of the total heat transferred across the membrane and 6% of the total heat is lost through the membrane [20,32]. Peclet number (Pe) can be used to determine the ratio of convective to conductive heat transfer rates within the membrane pores, and this dimensionless number can be represented as [12]

$$Pe = RePr = \frac{Nc_p}{h_m} \tag{5.13}$$

where N is the vapor flux and c_p is the heat capacity of the vapor.

Heat due to vapor permeation through the membrane:

$$Q_{mv} = N\Delta H_v = h_v\left(T_{fm} - T_{pm}\right) \tag{5.14}$$

where N and ΔH_v are the vapor flux through the membrane and the latent heat of vaporization of the volatile component, respectively. h_v, the heat transfer coefficient related to the water vapor flux, can be expressed as

$$h_v = \frac{N\Delta H_v}{\left(T_{fm} - T_{pm}\right)} \tag{5.15}$$

The total heat flux transferred through the membrane considering the contribution of both conduction and evaporation can be expressed as

$$Q_m = Q_{mc} + Q_{mv} = h_m \left(T_{fm} - T_{pm} \right) + h_v \left(T_{fm} - T_{pm} \right) = h_c \left(T_{fm} - T_{pm} \right)$$

$$(5.16)$$

where h_c is the effective membrane heat transfer coefficient and can be expressed as

$$h_c = h_m + h_v \tag{5.17}$$

5.7.5 Convective heat transfer in the permeate boundary layer

The convective heat flux, Q_p, from the membrane/permeate surface to the bulk permeate side across the permeate thermal boundary layer can be written as

$$Q_p = h_p \left(T_{pm} - T_{pb} \right) \tag{5.18}$$

where h_p is the heat transfer coefficient of the thermal boundary layer at the permeate side. As this heat transfer is associated with the temperature polarization effect, the membrane surface temperature (T_{pm}) is higher than the bulk permeate temperature (T_{pb}). This is true in all MD configurations except VMD due to a vacuum in the permeate side of the membrane.

At steady state conditions, the overall heat transfer flux through the MD system, Q, is given by

$$Q_f = Q_m = Q_p = Q \tag{5.19}$$

$$Q = h_f (T_{fb} - T_{fm}) = N\Delta H_v + \frac{k_m \left(T_{fm} - T_{pm} \right)}{\delta} = h_p \left(T_{pm} - T_{pb} \right) \quad (5.20)$$

5.7.6 Overall heat transfer coefficient (*U*) and temperature polarization coefficient (TPC)

Combining the equations for conductive and convective heat transfers, the total heat flux is expressed as

$$Q = \left[\frac{1}{h_f} + \frac{1}{h_c} + \frac{1}{h_p} \right]^{-1} \left(T_{fb} - T_{pb} \right) \tag{5.21}$$

The overall heat transfer coefficient, U, for the MD process can be written from Eq. (5.21) as

$$U = \left[\frac{1}{h_f} + \frac{1}{h_c} + \frac{1}{h_p} \right]^{-1} = \frac{Q}{\left(T_{fb} - T_{pb} \right)} \tag{5.22}$$

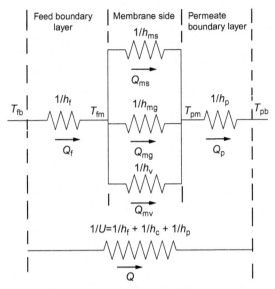

Figure 5.9 Heat transfer fluxes and resistances in MD.

The combined feed boundary layer, membrane, and permeate boundary layer resistance, $1/U$, is actually obtained as a series of three resistances in three regions as shown in Figure 5.9 named feed boundary layer resistance $(1/h_f)$, effective membrane resistance $(1/h_c)$, and permeate boundary layer resistance $(1/h_p)$.

Combining Eqs. (5.16), (5.18), and (5.19), at steady state conditions we can write

$$Q = h_c \left(T_{fm} - T_{pm} \right) \tag{5.23}$$

Similarly, Eqs. (5.7), (5.18), and (5.19) give

$$Q = h_f(T_{fb} - T_{fm}) = h_p \left(T_{pm} - T_{pb} \right) \tag{5.24}$$

From Eqs. (5.23) and (5.24), we can write

$$\frac{T_{fm} - T_{pm}}{T_{fb} - T_{pb}} = \frac{1}{1 + (h_c/h_f) + (h_c/h_p)} = \frac{h'}{h' + h_c} \tag{5.25}$$

where h' is the overall boundary layer heat transfer coefficient and is given by

$$\frac{1}{h'} = \frac{1}{h_f} + \frac{1}{h_p} \tag{5.26}$$

Comparing Eqs. (2.4) and (2.23), we have

$$\text{TPC} = \frac{1}{1 + (h_c/h_f) + (h_c/h_p)} = \frac{h'}{h' + h_c} = \frac{1}{1 + (h_c/h')} \tag{5.27}$$

To express the TPC in terms of overall heat transfer coefficient, combining the equations for U, h, and TPC, above, we can write

$$\text{TPC} = 1 - \frac{U}{h'} \tag{5.28}$$

5.7.7 Estimation of the interfacial temperatures

The two interfacial temperatures—(T_{fm}) at the feed/membrane surface and (T_{pm}) at the permeate/membrane surface as shown in Figure 5.9—cannot be measured directly, but can be calculated from the bulk temperatures and heat transfer coefficients [13,32,35].

At steady state, from the heat balance equations, T_{fm} and T_{pm} can be calculated from the following equations:

$$T_{fm} = \frac{h_m\{T_{pb} + (h_f/h_p)\,T_{fb}\} + h_f\,T_{fb} - N\Delta H_v}{h_m + h_f(1 + (h_m/h_p))} \tag{5.29}$$

$$T_{pm} = \frac{h_m\{T_{fb} + (h_p/h_f)\,T_{pb}\} + h_p\,T_{pb} + N\Delta H_v}{h_m + h_p(1 + (h_m/h_f))} \tag{5.30}$$

From these nonlinear equations the unknown values of T_{fm}, T_{pm}, and N can be estimated by applying an iteration method. This calculation requires estimation of the three heat transfer coefficients, h_f, h_m, and h_p. The membrane heat transfer coefficient, h_m, can be calculated from Eqs. (5.9) and (5.10). The boundary layer heat transfer coefficients, h_f and h_p, are almost always estimated from empirical correlations. Heat transfer coefficient in the boundary layers is affected by many factors such as the structure of the membrane module, different dimensions of the flow channel (such as equivalent diameter, length, etc.) in the module, flow velocities of the feed and permeate in the module, the physical properties (viscosity, density, thermal conductivity, heat capacity, etc.) of feed and permeate, the operation temperature, and so on. For turbulent flow inside circular tubes, a recognized empirical correlation, known as the Sieder–Tate equation, is popularly employed [12,22].

$$Nu = 0.023Re^{0.8}Pr^{1/3}\phi_\mu \tag{5.31}$$

where the dimensionless numbers are the Nusselt number, $Nu = hD/k$ (where D is the tube diameter, h is the tube side heat transfer coefficient, k is the thermal conductivity of the liquid), Reynolds number, $Re = DG/\mu$ (where G and μ are the mass velocity and bulk viscosity of the liquid, respectively), Prandtl number, $Pr = c_p\mu/k$ (where c_p is the liquid heat capacity), and the heating and cooling correction factor, $\phi_u = (\mu/\mu_w)^{0.14}$ (where μ_w is the liquid viscosity at the wall). In using this equation the physical properties of the liquid (Eq. (5.29) can also be used in case of gas/vapor), except for μ_w, are evaluated at the bulk temperature. Eq. (5.29) is recommended for Reynolds numbers above 6,000 and for tubes having a large ratio of tube length (L') to tube diameter (D), and should not be used for molten metals, which have abnormally low Prandtl numbers. It holds for $Re > 6,000$, a Pr between 0.7 and 16,000, and $L'/D > 60$. Near the tube entrance, where the thermal boundary layer is still forming, the local heat transfer coefficient is greater than h_α for fully developed flow, and the relation for short tubes, ($L'/D < 50$), is corrected by using the following equation:

$$\frac{h}{h_\alpha} = 1 + \left(\frac{D}{L'}\right)^{0.7} \text{ or } \frac{h}{h_\alpha} = 1.33 \left(\frac{D}{L'}\right)^{0.055} \qquad (5.32)$$

where h_α is calculated from Eq. (5.29). In case of a noncircular flow channel, the characteristic dimension D in the previous correlations is taken as an equivalent diameter d_h, which is defined as four times the hydraulic radius (r_h). The hydraulic radius is defined as the ratio of the cross-sectional area (S) of the flow channel to the length of the wetted perimeter (L) of the flow channel in contact with the fluid:

$$d_h = 4r_h = 4\frac{S}{L} \qquad (5.33)$$

For circular flow channels $d_h = D$; for square flow duct with a width of side w, $d_h = w$; and for parallel plates separated by distance w, $d_h = 2w$.

The heat transfer coefficients for the turbulent flow ($2,500 < Re < 1.25 \times 10^5$, and $0.6 < Pr < 100$) can be applied in a flat membrane module by the following equation [35]:

$$Nu = 0.023Re^{0.8}Pr^n \qquad (5.34)$$

where n is a constant, and equals 0.4 for heating and 0.3 for cooling, respectively.

For laminar flow ($Re < 2100$) in a circular tube, the above Sieder–Tate equation is successfully employed in the form of the following equation [37]:

$$Nu = 1.86(DRePr/L')^{1/3}\phi_u \qquad (5.35)$$

For laminar flow, a variety of mathematical correlations are applied in MD process. In case of $Re < 2,100$ in a flat membrane module, the following equation is used [38]:

$$Nu = 1.86\left(RePr\frac{d_h}{L}\right)^{0.33} \qquad (5.36)$$

where d_h and L are the hydraulic diameter and the chamber length of the flat module, respectively.

The heat transfer coefficients in laminar liquid flow for circular tubes with constant wall temperature are related to the Graetz number (Gz) in the form of the following equations [39]:

$$Nu = 3.66 + \frac{0.067\,Gz}{1 + 0.04\,Gz^{2/3}}, \quad Gz = \frac{mc_p}{kL'} \qquad (5.37)$$

where m is the mass flow rate, c_p is the liquid heat capacity, k is the liquid thermal conductivity, and L' is the length of the tubes. In the case of noncircular flow channels, Eq. (5.35) is not recommended for application.

In case of laminar flow in the hollow fiber membrane module, the heat transfer coefficient in the feed boundary layer of the feed solution circulating inside the fibers is expressed as [40]:

$$Nu = 1.615\left(\frac{d_h}{L}\right)^{1/3}(RePr)^{1/3} \qquad (5.38)$$

There are a variety of empirical correlations [40,41] applied for laminar and turbulent flow to estimate heat transfer coefficients for different geometries and heat transfer mechanisms, and may be useful in MD.

5.7.8 Evaporation efficiency

Because the MD process is based on evaporation to effect a desired separation, the amount of heat required in this phase change step is called efficient heat. The term Q_{mv} in Eq. (5.39) stands for this efficient heat contribution in the evaporation of the volatile component. One of the efficiency parameters in the MD process is evaporation efficiency (EE), which can be defined as the ratio of heat transfer through the membrane pores by vapor migration and the total heat transfer through the membrane [42]. Mathematically, the EE is expressed for well-insulated modules by

$$EE = \frac{Q_{mv}}{Q_{mv} + Q_{mc}} = \frac{N\Delta H_v}{Q} = \frac{N\Delta H_v}{N\Delta H_v + h_m \left(T_{fm} - T_{pm} \right)} \qquad (5.39)$$

This can be alternately written as

$$\frac{1}{EE} = 1 + \frac{\left(k_m / \delta \right) \left(T_{fm} - T_{pm} \right)}{\Delta H_v K_m \left(p_{fm} - p_{pm} \right)} \qquad (5.40)$$

where it can be seen that $\left(k_m / \delta \right)$ has a direct influence on EE.

Mathematically, the EE also is expressed for well-insulated modules by

$$EE = \frac{N\Delta H_v}{m_h C_p \left(T_{fb-in} - T_{fb-out} \right)} \qquad (5.41)$$

where m_h is the feed flow rate, C_p is the specific heat of the feed, T_{fb-in} and T_{fb-out} are the feed inlet and outlet temperatures, respectively, of the membrane module.

The evaporation efficiency can be maximized by reducing the internal heat loss by conduction through the membrane, the temperature polarization effect, and the external heat loss to the environment. In some studies, the maximum evaporation efficiency of 39% has been reported for the DCMD process [43] and about 90% for the VMD process.

5.8 MASS TRANSFER IN MEMBRANE DISTILLATION

Mass transfer in the MD process can be described in three steps: mass transfer in the feed side, mass transfer across the membrane, and mass transfer in the permeate side.

5.8.1 Mass transfer in the feed side

The mass transfer of the volatile components takes place from the bulk feed to the membrane surface. We can divide mass transfer phenomena in the feed side into three steps according to the number of components present in the feed.

5.8.2 A single volatile component feed

When the feed in MD is the pure solvent, the feed side membrane surface concentration is same as that at the bulk side; that is, no concentration gradient is built up. Therefore, there is no contribution of mass transfer resistance in the feed side; all mass transfer resistance is associated with the membrane itself.

5.8.3 Nonvolatile solute(s) with one volatile component

In the case of a liquid mixture containing a nonvolatile solute (B) in a volatile solvent (A), evaporation of the volatile component (A) at the membrane pore entrance results in the build-up of the nonvolatile component (B) near the membrane surface. That is, the concentration of the nonvolatile component (C_{Bm}) at the membrane surface is higher than that at the bulk feed (C_{Bb}). On the other hand, the volatile component's concentration (C_{Am}) at the membrane surface is less than that at the bulk feed (C_{Ab}). The region near the feed side membrane surface, where the concentration of nonvolatile and volatile component is achieved, is known as the concentration boundary layer or concentration polarization (CP) layer of thickness δ_{fc}, shown in Figure 5.10.

The resistance developed by the concentration boundary layer for the mass transfer of volatile component from bulk feed to membrane surface through the CP layer is referred to as concentration polarization. Therefore, concentration polarization reduces the transmembrane flux. The concentration of nonvolatile solute becomes higher and higher as the evaporation at the membrane surface is continued. As a result, supersaturation of the solute may be reached that affects the efficiency of the membrane process.

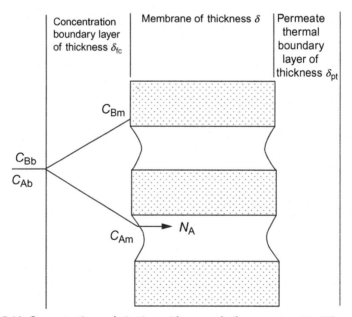

Figure 5.10 Concentration polarization with one volatile component in MD.

This supersaturating is not only reducing the transmembrane flux markedly due to reduction of an imposed driving force, but also starts deposition of a cake layer, resulting in fouling and scaling phenomena that increase the chance of membrane wetting and also add extra heat transfer resistance.

To characterize the concentration polarization—that is, to express the mass transfer resistance within the concentration boundary layer—the concentration polarization coefficient, CPC, is used. CPC is defined as the ratio of concentration of solute (B) at the membrane surface to that at the bulk feed and is represented as

$$CPC = \frac{C_{Bm}}{C_{Bb}} \tag{5.42}$$

The concentration polarization coefficient is defined as [44]

$$CPC = \frac{C_{Bm} - C_{Bb}}{C_{Bb}} \tag{5.42}$$

The resistance imposed by the concentration boundary layer is found to be less than that of thermal boundary layer. The thermal boundary layer resistance usually is evaluated as the temperature polarization coefficient (TPC). The effect of the TPC is observed to be more pronounced than the CPC. However, when feed contains more than one volatile component, the influence of the CPC on mass transfer has considerable effect. Like temperature polarization, the concentration polarization effect can be minimized by creating eddies and turbulence under high flow rates inside the flow channels, which enhances both heat and mass transfer. According to the theory of mass transfer in the boundary layer, the molar flux of the volatile component A through the feed side concentration boundary layer can be mathematically expressed from the Fick's law and mass balance equation [13]:

$$N_A = Ck_f^m \ln \frac{C_{Bm}}{C_{Bb}} = Ck_f^m \ln CPC \tag{5.43}$$

where N_A is the mass flux of volatile component A, C is the summation of concentration of A and nonvolatile component B in bulk feed, and k_f^m is the mass transfer coefficient of the volatile component (A) through the concentration boundary layer.

Temperature and concentration polarizations reduce the effective driving force, which can be measured by the vapor pressure polarization coefficient (VPC), defined as [44]

$$VPC = \frac{p_{fm} - p_{pm}}{p_{fb} - p_{pb}} \tag{5.44}$$

The vapor pressure difference has been taken into account for the real driving force for mass transport, and the coefficient VPC is introduced to measure the reduction of the imposed force $(p_{fb} - p_{pb})$ due to the existence of both temperature and concentration polarizations. When the feed is pure water, the VPC coincides approximately with TPC. In case of solutions of nonvolatile solutes, VPC differs from TPC because VPC depends on concentration of nonvolatile solutes and temperature. Therefore, VPC is the real representation of the reduction of driving force (that is flux). VPC generally depends on the temperature, solution concentration, and recirculation flow rate.

When the feed in the MD process is water or diluted aqueous solutions, and bulk temperature difference $(T_{fb} - T_{pb})$ is less than 10 °C, the following equation is often used in the MD literature [21]:

$$\frac{(p_{fm} - p_{pm})}{(T_{fm} - T_{pm})} = \frac{(p_{fb} - p_{pb})}{(T_{fb} - T_{pb})} \tag{5.45}$$

The boundary layer mass transfer coefficient, k_f^m, can be obtained experimentally or estimated by empirical equations in the following way.

According to the dimensional analysis, the general correlation is [45]

$$Sh = ZRe^{\alpha} Sc^{\beta} \tag{5.46}$$

in which

$$Sh = \frac{k_f^m d_h}{D_{AB}}, \quad Re = \frac{d_h v \rho}{\mu}, \quad \text{and} \quad Sc = \frac{\mu}{\rho D_{AB}} \tag{5.47}$$

where ρ is the liquid density, μ is the bulk liquid viscosity, d_h is the hydraulic diameter that is the characteristic diameter of the flow channel, D_{AB} is the binary diffusion coefficient in the liquid, v is the liquid velocity, Sh is the Sherwood number, and Sc is Schmidt number.

Equation (5.45) is the same form as Eq. (5.47), which can be used for calculations of heat transfer coefficients.

$$Nu = ZRe^{\alpha} Pr^{\beta} \tag{5.48}$$

where Nu, Re, and Pr are the Nusselt, Reynolds, and Prandtl numbers. Z, α and β are characteristic constants of the module design and liquid flow regime. These constants are considered equal in both equations, if the heat-mass transfer analogy is assumed. The constant Z is the parameter considering geometric characteristics and other conditions of the system. The other two constants α and β can be determined by the state of development of the velocity, temperature, and the concentration profile along the flow channel in the MD module [44]. The value of β is generally used as 0.33 when $Sc \gg 1$ and the operating conditions are that the length of the concentration profile entrance region is much larger than the length of the flow channel used in the MD module, whereas the value of α depends on flow regime. Schofileld et al. [46] used the Dittus–Boelter equation in the flat-sheet DCMD membrane module for the determination of the mass transfer coefficient for turbulent flow:

$$\frac{k_f^m d_h}{D_{AB}} = 0.023 Re^{0.8} Sc^{0.33} \tag{5.49}$$

To determine the mass transfer coefficient in the different MD literatures, the researchers applied an analogy between heat and mass transfer. Employing this analogy, the heat transfer equations can be used to determine the boundary layer mass transfer coefficients by replacing the Sherwood number for the Nusselt number, the Schmidt number for the Prandtl number, and the mass transfer Graetz number (Gz_M) for its heat transfer form [12]:

$$Sh = 0.023 Re^{0.8} Sc^{0.33} \phi_\mu \tag{5.50}$$

$$Sh = 3.66 + \frac{0.067 Gz_M}{1 + 0.04 Gz_M^{2/3}}, \quad Gz_M = \frac{m}{\rho D_{AB} L} \tag{5.51}$$

The solute concentration on the feed membrane surface C_{Bm} can be obtained from Eq. (5.42).

$$C_{Bm} = C_{Bb} \exp\left(\frac{N_A}{C k_f^m}\right) \tag{5.52}$$

For open flow channels (containing flat-sheet unsupported membrane) with or without spacer in the DCMD module the following equations are used [47,48]:

$$k_f^m = Z' D_{AB} \left(\frac{\rho v}{\mu}\right)^\alpha \left(\frac{\mu}{\rho D_{AB}}\right)^{0.33} \tag{5.53}$$

where the values of Z' and α in case of open flow channels in DCMD module are 8.22 and 0.45, respectively. On the other hand, the value of these constants for the flow channels with a spacer module are 0.90 and 0.72, respectively.

For mass transfer at the tube side of the hollow fiber membrane module in osmotic distillation process the following equations are used [40]. For the $Gz_M > 400$, the mass transfer coefficient is given by Leveque:

$$Sh = 1.615 \left(\frac{d_h}{L}\right)^{1/3} (ReSc)^{1/3} \qquad (5.54)$$

The correlation Eq. (5.54) is proposed by Viegas et al. [49] when $Re < 34$ and $Gz_M < 65$:

$$Sh = 0.2Re^{1.01} \left(\frac{d_h}{L} Sc\right)^{1/3} \qquad (5.55)$$

5.8.4 Two volatile component feed

The third case in which the second volatile component (R) continues evaporation at the feed side membrane surface results in the decrease in concentration of R across the concentration boundary layer of thickness δ_{fc}. Therefore, the concentration of component R decreases to C_{Rm} at the feed side membrane surface from C_{Rb} at the bulk feed phase as shown in Figure 5.11.

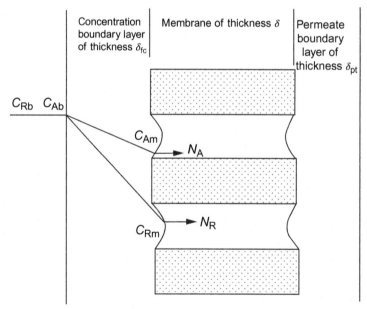

Figure 5.11 Concentration polarization with two volatile components in MD.

N_R, the mass flux of the volatile component R, through the concentration boundary layer (containing another volatile solvent, generally water, in the binary feed mixture) can be mathematically expressed [14] in the similar form of Eq. (5.43):

$$N_R = Ck_f^m \ln \frac{C_{Ab}}{C_{Am}} \qquad (5.56)$$

where C is the total molar concentration ($C_A + C_R$) in the liquid bulk phase and k_f^m is the mass transfer coefficient of the volatile component (R) across the CP layer in the feed side. The volatile component along with water is transported into the permeate side through the membrane in the MD process. Therefore, the separation efficiency is reduced. The separation efficiency or volatile component removal efficiency (S_e) of a volatile component in an aqueous mixture is determined as follows [50]:

$$S_e\% = \frac{C_0 - C_t}{C_0} \times 100 \qquad (5.57)$$

where C_0 and C_t are the volatile component concentrations at the start of the experiment and after time t, respectively.

The volatile component separation factor (S_t) is used to determine the degree of separation of the volatile component and can be measured by applying the following equation [51]:

$$S_t = \frac{(C_R / C_{H_2O})_p}{(C_R / C_{H_2O})_f} \qquad (5.58)$$

where C_R and C_{H_2O} are the concentrations of volatile and water at permeate side, p, and at the feed side, f, at time, t.

The high quality separation factor is achieved by selecting the proper membrane characteristics and good operating conditions that would favor the transport of volatile components over that water. Among the different MD configurations, vacuum membrane distillation (VMD) is applied for removal of volatile components present in aqueous solutions. In case of ammonia removal from aqueous solutions by the VMD process, experimental results showed that the separation efficiency is better at high feed temperatures, low pressure at the permeate side, and high initial feed concentrations and pH levels [52]. In this case, the pH plays a significant role. Higher feed temperature and lower permeate pressure enhance the ammonia removal efficiency and the transmembrane flux, but resulted in a lower ammonia separation factor. In the case of benzene removal from contaminated water by the VMD process, the overall mass transfer coefficient is approximately equal to the contribution from the feed liquid layer resistance neglecting the membrane resistance and the permeate side resistance. The overall mass transfer coefficient is considered to be the sum of the feed side boundary layer resistance and the membrane resistance.

The driving force for transport of the vapor through the feed side boundary layers is the vapor pressure difference across the boundary layers. Therefore, the resistance R_f to the vapor transfer due to temperature and concentration polarizations in the feed side is given by:

$$R_f = \frac{1}{K_f} = \frac{p_{fb} - p_{fm}}{N} \tag{5.59}$$

where p_{fb} and p_{fm} are the vapor pressures at the bulk feed and feed vapor-liquid interface, respectively. N and K_f are the vapor flux and the feed side transport coefficient, respectively. $1/K_f$ may be defined as transport resistance in which both concentration and temperature polarizations in the feed side have been taken into account. K_f is different from the feed side mass transfer coefficient k_f in both units and physical meaning. The effect of concentration polarization is taken into account during calculation of k_f, without consideration of temperature polarization.

5.8.4 Mass transfer in the permeate side

As almost 100% rejection of nonvolatile components at the feed side in the MD process is observed and as long as no wetting of membrane occurs, no concentration boundary layer is formed in the permeate side. Therefore, the resistance to mass transfer due to concentration polarization in the permeate side is believed to be zero.

In the permeate side, the transport resistance R_p may be written as:

$$R_p = \frac{1}{K_p} = \frac{p_{pm} - p_{pb}}{N} \tag{5.60}$$

where p_{pm} and p_{pb} are the vapor pressures at the permeate vapor-liquid interface and the bulk permeate, respectively. $1/K_p$ is the permeate side transport resistance in which only temperature polarization is incorporated. Concentration polarization is not taken into account for the calculation of permeate side transport coefficient (K_p), or $1/K_p$ because no concentration boundary layer is developed in the permeate side.

Permeate side resistance can be classified into four steps depending on the method by which the permeating vapor is recovered from the pores at the permeate side.

5.8.5 Permeate side resistance in AGMD

Permeate side total resistance R_p is offered by four domains, as shown in Figure 5.12. These are air/vapor gap (g), condensate film (f), cooling plate (p), and the cold fluid (c).

A mass transfer resistance, R'_{Mi}, at any domain i can be expressed as [53]:

$$R'_{Mi} = \frac{\Delta p_i}{N} \tag{5.61}$$

where Δp_i is vapor pressure difference across the i-th domain.

The mass transfer resistance in the condensate film R'_{Mf} is very small compared to the resistance of the air/vapor gap domain. Therefore, the mass transfer resistance in the permeate side is totally offered by the air/vapor gap domain and is represented via a relationship:

$$R'_{Mg} = \frac{\Delta p_g}{N} = \frac{p_{gm} - p_{gf}}{N} \tag{5.62}$$

where $\Delta p_g = p_{gm} - p_{gf}$ is vapor pressure difference across the domain air/vapor gap. AGMD flux and conduction heat transfer decrease with the increase of air/vapor gap width. The air gap width (1–5 mm) have the strongest effect on the air/vapor gap domain mass transfer resistance (R'_{Mg}).

5.8.6 Permeate side resistance in DCMD

In Figure 5.12, the domains g, f, and p are eliminated for DCMD, and as a result the cold permeate is in direct contact with the cold side of the membrane. The permeate side cold liquid mass transfer resistance, R''_{Mc}, can be evaluated by

$$R''_{Mc} = \frac{\Delta p_c}{N} = \frac{p_{pm} - p_{pb}}{N} \tag{5.63}$$

where $\Delta p_c = p_{pm} - p_{pb}$ is the vapor pressure difference across the cold permeate, and p_{pm} and p_{pb} are the vapor pressures at the permeate side of the vapor–liquid interface and bulk permeate, respectively.

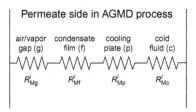

Figure 5.12 Permeate side mass transfer resistances in the AGMD process.

5.8.7 Permeate side resistance in SGMD

To reduce permeate side mass transfer resistance in AGMD across the stagnant air-vapor film, the SGMD configuration is adopted, in which permeate side resistance is reduced by the air blowing over the membrane surface. The permeate flux in SGMD depends on the sweep air velocity and air temperature at the module inlet [26]. SGMD flux increases through a maximum as the gas velocity increases, but then begins to decrease. The permeate side mass transfer boundary layer thickness reduces (i.e., reduction of permeate mass transfer resistance) with an increase in the gas velocity, giving rise to an increase of flux. However, the pressure of the sweep gas increases as a consequence of increasing the gas velocity, and the permeate side boundary layer resistance increases. As a result, the curve (flux vs. sweep gas velocity) shows an optimum.

5.8.8 Permeate side resistance in VMD

In case of the removal of a volatile component from an aqueous mixture, the mass transfer resistance at the permeate side, R_{Mg}''', can be written as [27]

$$R_{Mg}''' = \frac{c_l}{k_g' H_a} \tag{5.64}$$

where c_l is the liquid concentration, k_g' is the permeate side gas mass transfer coefficient, and H_a is the Henry's law constant.

The volatile component diffusion coefficient in water is many fold times slower (e.g., 10,000 times slower in the case of benzene) than it will be in air. Therefore, the calculated value of the permeate side gas phase resistance according to Eq. (5.63) is very small in comparison with the feed side resistance. The value of R_{Mg}''' is further reduced due to an increase of the gas diffusion coefficient because very low pressure is maintained by applying a vacuum at the downstream side in the VMD process. The mass transfer resistance at the permeate side is thus neglected.

Therefore, the resistance to mass transfer in VMD is located in the feed side and within the membrane pores. The overall mass transfer resistance $(1/K_{OV})$ in VMD modules can be represented as the sum of the feed side boundary layer resistance $(1/K_f)$ and the membrane resistance $(1/K_m)$ [54]:

$$\frac{1}{K_{ov}} = \frac{1}{K_f} + \frac{1}{K_m} \tag{5.65}$$

where K_f and K_m are feed and membrane transport coefficients, respectfully. This equation was found to apply well to ammonia removal from an aqueous solution [52]. The overall mass transfer coefficient in this system is approximately equal to the contribution to the liquid layer resistance.

5.8.9 Mass transfer through the membrane pores

Like Darcy's law for laminar flow in packed beds, the MD flux N through the dry porous membrane is proportional to the vapor pressure difference across the membrane, $(p_{fm} - p_{pm})$, and can be expressed as

$$N = K_m (p_{fm} - p_{pm}) \tag{5.66}$$

where K_m is the membrane distillation coefficient or permeability of the membrane or membrane transport coefficient, and p_{fm} and p_{pm} are vapor pressures of the transporting volatile component at the feed and permeate vapor–liquid interfaces, respectively.

The value of K_m depends on the temperature, pressure, and composition within the membrane as well as the membrane structure. The vapor pressures of the volatile transporting components p_{fm} and p_{pm}, at the temperatures T_{fm} and T_{pm}, respectively, is a function of temperature and is related to the activity coefficient of the liquid solution (nonvolatile solute with one volatile component) by

$$p_{im} = (1 - x_{im}) a_{im} p_{im}^o, \quad i = f, p \tag{5.67}$$

where a_{im} and x_{im} are the interfacial activity coefficient of the volatile component and interfacial nonvolatile solute component mole fraction in the liquid solution, respectively; f and p stand for feed and permeate sides, respectively. p_{im}^o is the interfacial vapor pressure of pure volatile liquid and can be calculated using the Antoine equation by

$$p_{im}^o = \exp\left(23.238 - \frac{3{,}841}{T_{im} - 45}\right) \tag{5.68}$$

where p_{im}^o is pure volatile liquid interfacial vapor pressure in Pascal and T_{im} is the corresponding interfacial temperature in Kelvin. The effect on the p_{im}^o is given by the Kelvin equation [12]:

$$p_{im}^{o\prime} = p_{im}^o \exp\left(\frac{2\gamma_1}{r' cRT}\right) \tag{5.69}$$

where $p_{im}^{o\prime}$ is the pure volatile liquid saturation pressure above a convex liquid surface of curvature r', p_{im}^o is the pure volatile liquid saturation pressure above a flat surface, γ_1 is the liquid surface tension, c is liquid molar density,

R is the gas constant, and T is the temperature. Most of the MD papers neglect the curvature effect on the pressure since its effect is small.

The feed solutions in the case of MD systems can be considered very dilute solutions. As a result, the water vapor pressure on each side of the membrane will be equal to the saturation pressure of the pure water— $p_{im} = p_{im}^o$ —and the Clausius–Clapeyron equation can be used to simplify the vapor pressure temperature relation as shown in Eq. (5.70):

$$\frac{\Delta p^o}{\Delta T} = \frac{dp^o}{dT} = \frac{\Delta p_{im}^o}{\Delta T_m} = \frac{p_{fm}^o - p_{pm}^o}{T_{fm} - T_{pm}} = \frac{p^o \Delta H_v'}{RT^2} \tag{5.70}$$

where p^o is the average values of the saturation pressure of water at average absolute temperature T within the membrane, and $\Delta H_v'$ is the molar heat of vaporization at temperature T.

The Clausius-Clapeyron equation can be applied to Eq. (5.70) to obtain the following equation in case of very dilute aqueous solutions:

$$N = K_m \frac{\left(p_{fm}^o - p_{pm}^o\right)}{\left(T_{fm} - T_{pm}\right)}\left(T_{fm} - T_{pm}\right) = K_m \frac{p^o \Delta H_v'}{RT^2}\Delta T_m \tag{5.71}$$

The value of a, which is a function of temperature and composition, can be evaluated experimentally or by applying any one of the different kinds of equations such as NRTL, UNIQUAC, Wilson, and van Laar. According to the Raoult's Law, the vapor pressure–composition relation can be written in case of ideal dilute solutions (nonvolatile solute with one volatile component):

$$p_{im} = (1 - x_{im})p_{im}^o \tag{5.72}$$

where x_{im} is the mole fraction of the nonvolatile solute at the interface. For the mixture of water and NaCl, the activity coefficient of water can be calculated by the following equation:

$$a_{water} = 1 - 0.5x_{NaCl} - 10x^2 \tag{5.73}$$

In case of nonideal solutions, the effect of nonvolatile solute in the feed (nonvolatile solute with one volatile component) is to reduce the vapor pressure, which is represented by Eq. (5.72), which describes the deviation from Raoult's Law. From Eq. (5.43), the interfacial mole fraction of the solute, x_{im}, can be calculated by

$$x_{im} = x_b \exp\left(\frac{N}{k_f^m \rho}\right) \tag{5.74}$$

where x_b is the mole fraction of nonvolatile component in the bulk; x_{im} is at the membrane surface; N is the mass transfer flux across the membrane; ρ is the density of the solution; and k_f^m is the mass transfer coefficient of the volatile component through the concentration boundary layer. The value of k_f^m is usually estimated from the analogy between heat and mass transfers, which was described earlier.

The hydrophobic membrane used in MD is a porous media. The two models—Restricted Gas model and Dusty Gas model (DGM)—are used for gas transport through porous media. The DGM normally is used for describing mass transport through the membrane during membrane distillation. The DGM is derived by applying kinetic theory of gases to the interaction of both gas–gas and gas–solid molecules, with the porous media treated as "dust" in the gas. The DGM describes mass transport in porous media by four possible mechanisms [55]. According to the DGM, there are the four possible mass transfer resistances in MD using an electrical analog as illustrated in Figure 5.13.

Viscous or momentum or Poiseuille flow resistance results from transfer of momentum to the supported membrane, molecular resistance comes from collisions of a diffusing molecule with other molecules, or collisions of a transporting molecule with the walls of the membrane pores gives Knudsen resistance. The fourth mode of diffusion in the DGM model is the surface flow or diffusion, in which molecules move along a solid surface in an adsorbed layer. The surface flow mechanism is assumed independent of the other three mechanisms and is neglected in MD modeling because it has

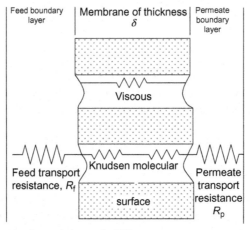

Figure 5.13 Mass transfer resistances in MD.

little influence on the whole process due to the weak molecule membrane interaction. It can be incorporated in the development of hydrophilic MD membranes.

The mechanism of flow of gas molecules through porous membranes depends upon the mean pore diameter of the membrane and mean free path of the gas molecules. The mean free path can be defined as the average distance the molecule travels between successive collisions. The Knudsen number (K^n) is used to regulate mass transfer mechanism across the membrane with nonuniform pore sizes. The Knudsen number (K^n) is defined as the ratio of the mean free path (λ) to the mean pore diameter (d_p); that is, $K^n = \lambda/d_p$. Considering the binary mixture (e.g., water vapor and air), the mean free path of one component as represented 1 (e.g., water vapor) in the other component as represented 2 (e.g., air) is evaluated at the average membrane temperature (T) from the kinetic theory [12,56,57]:

$$\lambda_{1-2} = \frac{k_B T}{P_t \pi ((\sigma_1 + \sigma_2)/2)^2 \sqrt{1 + (M_1/M_2)}} \tag{5.75}$$

where k_B is the Boltzman constant $(1.381 \times 10^{-23}$ J/K); P_t is the total pressure inside the membrane; σ_1 and σ_2 are the collision diameters for component 1 $(2.641 \times 10^{-10}$ m for water vapor) and for component 2 $(3.711 \times 10^{-10}$ m for air), respectively; and M_1 and M_2 are the molecular weights of components 1 and 2, respectively. At the average membrane temperature of 60 °C and at 1.013×10^5 Pa or 1 atm for DCMD, the mean free path of water in air (λ_{1-2}) in DCMD system is 0.11 μm [41].

In accordance with the Knudsen number (K^n), various kinds of mechanisms have been proposed for transport of gases or vapors through porous membranes: the Knudsen flow model, viscous flow model, ordinary diffusion model, and a combination thereof.

5.8.10 Knudsen flow or free molecule flow

If the mean free path of transporting gas molecules is larger than pore diameter (i.e., $K^n > 1$ or $d_p < \lambda$) or the gas or vapor density is so low, the molecule–pore wall collisions dominate over the molecule–molecule collisions in the mass transfer process, and the mass transfer within the membrane pores is regulated by the Knudsen flow, which can be represented by K. The resistances coming from the viscous flow or Poiseuille flow (P) and molecular diffusion (M) of the gas through the porous membrane can be neglected. The mass flux $(kg/(m^2 sec))$ from the Knudsen diffusion model [36,41] can be expressed as

$$N_K = \frac{2}{3} \frac{\varepsilon r}{\tau \delta} \left(\frac{8M}{\pi R T} \right)^{1/2} \left(p_{fm} - p_{pm} \right) \tag{5.76}$$

where ε, r, τ, and δ are the porosity, pore radius, tortuosity, and thickness of the porous hydrophobic membrane, respectively; M is the molecular weight of the transporting vapor and T is the average membrane surface absolute temperature inside the pores, $(T_{fm} + T_{pm})/2$.

Equation (5.76) can be also expressed in terms of the Knudsen diffusion coefficient (D_K) corrected for membrane porosity (ε) and pore tortuosity (τ), and the concentration driving force $\left(c_{fm} - c_{pm} \right)$.

$$N_K = D_K \frac{\varepsilon}{\tau \delta} \left(c_{fm} - c_{pm} \right) \tag{5.77}$$

where c_{fm} and c_{pm} are the concentrations of the transporting component at the feed and permeate vapor–liquid interfaces, respectively, and

$$D_K = \frac{2}{3} r \left(\frac{8RT}{\pi M} \right)^{1/2} \quad \text{and} \quad c_{im} = \frac{p_{im} M}{RT} \tag{5.78}$$

The term $(8RT/\pi M)^{1/2}$ is the mean molecular speed.

In the VMD process, the small pore size membranes are employed for avoiding pore wetting and the low downstream pressure is maintained. In such a situation, the diffusion process can be described by the Knudsen diffusion, and the flux can be expressed as:

$$N_K = \frac{2}{3} \frac{\varepsilon r}{\tau} \left(\frac{8M}{\pi R T} \right)^{1/2} \nabla p_{im} \tag{5.79}$$

where ∇p_{im} is the pressure gradient of the transporting component between the membrane surfaces and can be described by

$$\nabla p_{im} = \frac{\Delta p_{im}}{\delta} = \frac{\left(p_{fm} - p_{pm} \right)}{\delta} \tag{5.80}$$

where Δp_{im} is the pressure difference of the transporting component between the membrane surfaces.

5.8.11 Viscous or convective, or bulk or Poiseuille flow

If the pore size is much larger than the mean free path of the transporting molecule (i.e., $K^n < 0.01$ or $\lambda < 0.01 d_p$), the molecule–wall collision can be neglected, and the resistance resulting from the gas viscosity. In this case, the gas acts as a continuous fluid driven by a pressure gradient and by molecule–molecule collisions. The mass transfer regulated in this way is known as viscous or the Poiseuille flow mechanism. The mass flux $(kg/(m^2 \; sec))$ of the transporting species through the pores of the membrane by the Poiseuille flow mechanism can be written as [58]:

$$N_P = 0.125 \frac{\varepsilon r^2}{\tau \delta} \left(\frac{M p_m}{\mu R T} \right) \left(p_{fm} - p_{pm} \right) \qquad (5.81)$$

where p_m is the vapor pressure of the transporting volatile component in the membrane and μ is the viscosity of vapor–air mixture within the membrane at temperature T.

In the DCMD process, the hot feed and the cold permeate are brought into contact with the membrane under atmospheric pressure, the total pressure (i.e., ≈ 1 atm.) is almost constant, and therefore, the viscous flow can be neglected. For mass transfer of single component vapor or gas across the membrane, the mass transfer of the component by molecular diffusion mechanism within the pores is neglected. The driving force for this mass transfer is pressure gradient and the mechanism of mass transfer is Poiseuille flow.

5.8.12 Ordinary (continuum) or molecular diffusion

If $K'' < 0.01$ or $\lambda < 0.01 d_p$, molecule–molecule collisions are dominant over molecule–pore wall collisions so that mass transfer of the transporting component is governed by molecular diffusion. In molecular diffusion, the different species of a mixture move relative to each other under the influence of concentration gradients. Molecular diffusion is used to describe the mass transfer of the vapor through the stagnant air trapped within the membrane pores due to the low solubility of air in water.

Schofield et al. [32] and Ding et al. [58] described the mass flux through a stationary film of air trapped within the membrane pore, and the mass flux $(kg/(m^2 \; sec))$ can be represented by applying the molecular diffusion model [32,58]:

$$N_M = \frac{1}{y_{ln}} \frac{D_{wa} \varepsilon}{\tau \delta} \frac{M}{RT} \left(p_{fm} - p_{pm} \right) \qquad (5.82)$$

where D_{wa} is the diffusivity of vapor through the air and y_{ln} is the log mean of the air mole fractions at the feed and permeate vapor–liquid interfaces.

Mass flux $(kg/(m^2 \; sec))$ for molecular diffusion in the DCMD process can be expressed as [35,36]:

$$N_M = \frac{\varepsilon}{\tau \delta} \frac{P_t D_{wa}}{P_a} \frac{M}{RT} \left(p_{fm} - p_{pm} \right) \qquad (5.83)$$

where P_t is the total pressure inside the pore assumed constant and equal to the sum of the partial pressures of air and vapor. P_a is the partial pressure of the air entrapped in the pores, which is generally taken as $|P_a|_{ln}$ (the log mean of the air pressures at the feed and permeate vapor–liquid interfaces). The value of quantity P_tD_{wa} (Pa m^2/s) for water–air can be calculated from the following expression [59]:

$$P_tD_{wa} = \left(1.895 \times 10^{-5}\right) T^{2.072} \tag{5.84}$$

Value of quantity P_tD_{wa} for water–air in Pa m^2/sec can be estimated using the relation:

$$P_tD_{wa} = \left(4.46 \times 10^{-6}\right) T^{2.334} \tag{5.85}$$

The solubility of air in water is so low that the air flux through the liquid boundary layers or the membrane pores can be neglected; therefore, the air can be treated as a stagnant film. According to the DGM, the flux (kg/(m^2 sec)) of the transporting component through this stagnant film can be written as

$$N_M = \frac{\varepsilon P D_{wa}}{\tau T^b} \frac{T^{b-1}}{|P_a|_{ln}} \frac{M}{R\delta} \left(p_{fm} - p_{pm}\right) \tag{5.86}$$

where Eqs. (5.83) and (5.86) are similar equations; the only difference is that the temperature has been separated into two terms, T^{b-1} and T^b, so that the value of $\varepsilon P D_{wa}/(\tau T^b)$ does not depend on temperature and pressure.

5.8.13 The Knudsen-molecular diffusion transition

When $0.01 < K^n < 1$ or $\lambda < d_p < 100\ \lambda$, collision between molecules is as important as the collision between molecules and the pore wall. The mass transfer equation for the transition region is based on the momentum balance including momentum transferred by gas molecules to other molecules as well as to the pore wall. In this case, mass transport takes place via the combined Knudsen/ordinary or molecular diffusion mechanism. The combination of Knudsen and molecular diffusion resistances in series is used to describe mass transfer for this region. If membranes have smaller air-filled pores (approximately less than 0.5 μm), molecule–pore wall collisions start as frequently as molecule–molecule collisions, and the Knudsen diffusion becomes important with molecular diffusion. The total flux (kg/(m^2 sec)) (N_{M-K}) is related to the Knudsen diffusion (N_K) and the molecular diffusion (N_M):

$$\frac{1}{N_{M-K}} = \frac{1}{N_K} + \frac{1}{N_M} \tag{5.87}$$

Combining Eq. (5.79) with Eqs. (5.86) and (5.87), finally Eq. (5.88) is obtained for calculation of the total mass flux (kg/(m^2 sec)) (N_{M-K}) in DCMD process.

$$N_{M-K} = \left[\frac{3\tau\delta}{2\,\varepsilon r} \left(\frac{\pi RT}{8M} \right)^{1/2} + \frac{\tau\delta}{\varepsilon} \frac{P_a}{P_t D_{wa}} \frac{RT}{M} \right]^{-1} \left(p_{fm} - p_{pm} \right) \qquad (5.88)$$

In the SGMD, the combined Knudsen–molecular diffusive flux is responsible for the transport processes and the flux can be written in the form of Eq. (5.89):

$$N_{M-K} = \frac{\varepsilon}{\tau\delta} \frac{M}{RT} \left(\frac{1}{D_K} + \frac{P_a}{P_t D_{wa}} \right)^{-1}_{ln} \left(p_{fm} - p_{pm} \right) \qquad (5.89)$$

where D_K is the Knudsen diffusion coefficient. The subindex indicates the logarithmic mean of $(1/D_K + P_a/PD_{wa})^{-1}$ in both membrane surfaces, where different values of P_a exist.

These equations are in the form of $N_{M-K} = K_m \left(p_{fm} - p_{pm} \right)$, and it is difficult to determine the value K_m analytically. Therefore, the value of K_m is determined from experimental data and then compared to the theoretical value given by the previous equations to determine the relative importance of Knudsen and molecular diffusion in mass transfer mechanism.

5.8.14 The Knudsen–Poiseuille transition

This type of transition mechanism may be observed under the following conditions.

5.8.14.1 The Knudsen–Poiseuille transition for single species MD system

When $0.01 < K^n < 1$ or $\lambda < d_p < 100\,\lambda$ and in case of mass transfer of a single species (e.g., water vapor) under imposed pressure gradient on the membrane, the resistance caused by molecular–molecular collision can be neglected, and in this case both Knudsen diffusion and Poiseuille flow have a noticeable contribution to mass transfer instead of predominating any one of Knudsen diffusion or Poiseuille flow. Therefore, using a combination of these two mechanisms, called the Knudsen diffusion–Poiseuille flow transition, is recommended to describe mass transfer. The total mass flux (kg/(m^2 sec)) (N_{K-P}) can be expressed as:

$$N_{K-P} = N_K + N_P \qquad (5.90)$$

where N_K and N_P represent the contribution of Knudsen diffusion and Poiseuille flow to mass transfer, respectively.

The flux equation can be transformed into the following equation for flux $(kg/(m^2 sec))$ by replacing N_K and N_P:

$$N_{K-P} = \left[\frac{2}{3} \frac{\varepsilon r}{\tau \delta} \left(\frac{8M}{\pi RT} \right)^{1/2} + \frac{\varepsilon r^2}{\tau \delta} \frac{p_m M}{8\mu RT} \right] \Delta p \qquad (5.91)$$

where Δp (Pa) is the pressure difference across the membrane pores, R (J/mol K) is the universal gas constant, M (kg/kmol) is the molecular weight of the species, T (K) is the temperature of the gas, μ (Pa sec) is the gas viscosity, and p_m is the average pressure within the membrane pores.

5.8.14.2 The Knudsen–Poiseuille transition in DCMD and VMD systems

The membrane permeability in DCMD can be increased by removing the stagnant air from the membrane pores by degassing the feed and permeate. Due to removal of air from the membrane pores, the two effects are noticed. First, since the air partial pressure in the membrane pores is very low, molecule–wall collision in mass transfer predominates over molecule–molecule collision. Therefore, the Knudsen diffusion resistance is the dominant resistance over the molecular diffusion resistance, which is in series with the Knudsen diffusion resistance in the electrical analog. Second, more or less air within the membrane results in an inability to maintain a constant total pressure across the membrane; the resistance coming from the gas viscosity increases with the increase of the driving force for viscous flux. Therefore, the Knudsen diffusion–Poiseuille flow transition mechanism is observed in the degassed DCMD systems.

In the VMD, only a trace amount of air exists in the membrane pores. Therefore, the mass transfer resistance caused by molecule–molecule collision can be neglected, and fluxes of VMD systems can also fall within the Knudsen diffusion–Poiseuille flow transition region.

According to the DCM, the flux $(kg/(m^2 sec))$ (N_{K-P}) in degassed DCMD or VMD systems can be expressed as:

$$N_{K-P} = \left[\frac{2}{3} \frac{\varepsilon r}{\tau \delta} \left(\frac{8M}{\pi RT} \right)^{1/2} \left(p_{fm} - p_{pm} \right) + \frac{\varepsilon r^2}{\tau \delta} \frac{PM}{8\mu RT} \left(P_{fm} - P_{pm} \right) \right] \qquad (5.92)$$

where M is the molecular weight of the transporting volatile component; $\left(p_{fm} - p_{pm} \right)$ (i.e., Δp_i) is the vapor pressure difference of transporting volatile component i across the membrane; $\left(P_{fm} - P_{pm} \right)$ (i.e., ΔP) is the total pressure difference across the membrane; and μ is the gas viscosity (pressure independent).

A more empirical model for the Knudsen–Poiseuille transition has been developed [60]. In case of fully Poiseuille flow, $N \propto P \Delta P$ and for fully Knudsen flux $N \propto \Delta P$, they correlated the experimental data and their model:

$$N = d' p_w^c \Delta p_w \qquad (5.93)$$

where d' is the membrane permeability constant, c is the exponent K–P transition, p_w is the water vapor pressure, and Δp_w is the water vapor pressure drop across the membrane. c can range from 0.0 for fully Knudsen flux to 1.0 for fully Poiseuille flow.

5.8.14.3 The molecular-Poiseuille transition

When $K^n < 0.01$ and total pressure gradient is imposed across the membrane, Poiseuille flow makes a noticeable contribution to mass transfer. Moreover, if the concentration gradient of the transporting component is built up across the membrane, molecular diffusion also makes a contribution to mass transfer. Under these conditions, the mass transfer is regulated by the molecular diffusion–Poiseuille flow transition mechanism, the total mass flux (N_{M-P}) is the sum of contributions of molecular diffusion (N_M), and Poiseuille flow (N_P):

$$N_{M-P} = N_M + N_P \qquad (5.94)$$

5.8.14.4 The Knudsen–molecular–Poiseuille transition

When $0.01 < K^n < 1$ and the total pressure gradient is imposed across the membrane, all the three basic mechanisms contribute to mass transfer. Therefore, the total mass flux (N_{K-M-P}) is the sum of contributions of Knudsen diffusion–molecular diffusion (N_{K-M}) and Poiseuille flow (N_P) [13].

$$N_{K-M-P} = N_{K-M} + N_P \qquad (5.95)$$

5.8.14.5 The membrane distillation coefficient versus temperature, membrane pore size, and transport mechanisms

The Knudsen–molecular diffusion–Poiseuille flow-transition (KMPT) model of Ding et al. [58] emphasizes the effects of temperature and membrane pore size on the value of MDC. It is observed that for membranes having a small pore diameter (0.1 μm), the MDC value changes very little with feed temperature, indicating that the mass transfer mechanism is regulated by the Knudsen–molecular diffusion with negligible contribution of the Poiseuille mechanism. It is found that with increase of pore size, the value of MDC increases with increase of feed temperature, indicating the importance of the Poiseuille flow mechanism. Under these conditions, the mass transfer mechanism is regulated by Poiseuille flow. The KMPT model says that the membrane with the larger pore size is more favorable to Poiseuille flow and less favorable to diffusion.

According to the Knudsen diffusion model, Poiseuille flow model, and the molecular diffusion model) [32], MDC depends slightly on a temperature decrease by about 3% with a 10 °C increase in average temperature. MDC will be strongly dependent on the membrane geometry, if convective transport (Poiseuille flow) is dominant. If diffusive transport predominates, the controlling parameter will be the average mole fraction of air present within the pores. Figure 5.14 illustrates mass transfer resistances under different transport regimes of diffusion, Knudsen and Poiseuille models.

5.8.15 Determination of membrane characteristics: Gas permeation (GP) test

The important membrane characteristics, such as porosity (ε), thickness (δ), and mean pore size (r), are important, and their values can be measured by gas permeation (GP) experiments of a pure single gas such as nitrogen. Unfortunately, water or other vapors in an MD system cannot be used directly to measure these values due to problems caused by condensation. The experiment materials involve the porous membrane to be investigated and nitrogen. The experiment is carried out by passing nitrogen through the porous membrane driven by a total pressure difference exerted between the two sides of the membrane. Under these conditions, the permeation of the single gas is regulated by Knudsen diffusion and Poiseuille flow mechanism. The steady state gas permeation flux is represented by Eq. (5.96). This equation can be written in the form

$$J_{K-P} = A_o + B_o p_m \tag{5.96}$$

where $J_{K-P} = \frac{N_{K-P}}{\Delta p}$, $A_o = \frac{2}{3} \frac{\varepsilon r}{\tau \delta} \left(\frac{8M}{\pi RT} \right)^{1/2}$ and $B_o = \frac{\varepsilon r^2}{\tau \delta} \frac{M}{8 \mu RT}$ (Pa) is the average pressure within the membrane pores. In order to get A_o and B_o, a gas permeation experiment is carried out at various p_m, while keeping pressure difference across the membrane constant. Under this condition, the gas permeation fluxes N_{K-P} (kg/m^2 sec) through the membrane are measured. According to Eq. (5.95), a straight line curve of J_{K-P} (kg/(m^2 sec Pa)) with

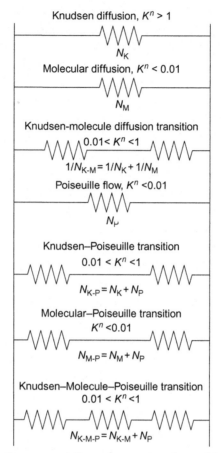

Figure 5.14 Schematic representation of mass transfer resistances within porous membrane and Knudsen number.

p_m is plotted. From the graph, the intercept with the y axis is A_o and the slope is B_o. The membrane characteristics can be obtained from A_o and B_o by using the following equations:

$$\frac{\varepsilon r}{\tau\delta} = \frac{3}{8}A_o\sqrt{\frac{2\pi RT}{M}}, \quad \frac{\varepsilon r^2}{\tau\delta} = 8B_o\frac{\mu RT}{M} \tag{5.97}$$

$$r = \frac{16}{3}\frac{B_o}{A_o}\sqrt{\frac{8RT}{\pi M}}\mu, \quad \frac{\varepsilon}{\tau\delta} = \frac{8\mu RTB_o}{Mr^2} \tag{5.98}$$

In the GP experiment, nitrogen is introduced from the cylinder pump into the membrane module, in which the flat sheet membrane is supported on a porous sintered body made of stainless steel or ceramic. The pressures at two sides of the membrane are controlled by valves, and measured with pressure sensors. Flow rate of the gas through the membrane is measured with a soap flow meter. A buffer tank used for getting a stable gas stream is kept between the membrane module and the soap flow meter.

The GP experimental study is carried out [13] by using a PTFE membrane under a constant pressure difference of 100,368 Pa. The drawn curve of J_{K-P} (kmol/(m^2 sec Pa)) with p_m is a straight line well correlated with the collected data. From the intercept (B_o) and the slope (A_o) values, the membrane characteristics can be calculated from Eqs. (5.97) and (5.98). The membrane characteristics vary slightly with pressure difference. During the gas permeation test, membrane thickness (δ) is not a constant and varies with pressure difference as described by Eq. (5.99). The reason may be due to the compaction of membrane during the experiment. This compaction of the membrane not only decreases membrane thickness with an increase pressure difference, but also leads to a change of the porosity and pore size. According to the equation proposed by Lawson and Lloyd [12] effective membrane thickness (δ_e) decreases linearly as the pressure increases (ΔP):

$$\delta_e = \delta_o + \delta' \Delta P \tag{5.99}$$

where δ_o is the uncompacted membrane thickness and δ' is the compaction parameter related to the membrane structure. Membrane permeability increases upon compaction due to the positive value of δ'. The effect of decrease in the diffusional path length through the membrane leading to an increase in the permeability outweighs the effects of a decrease in the membrane porosity and pore sizes, leading to a decrease in the permeability.

5.9 SOLAR-DRIVEN MEMBRANE DISTILLATION IN ARSENIC REMOVAL

5.9.1 Introduction

A solar-driven membrane distillation (SDMD) system as illustrated in Figure 5.15 has been used successfully in arsenic removal from groundwater [31]. In a solar-driven direct contact membrane distillation (DCMD) system with provision for flash vaporization, 100% arsenic removal is possible with a reasonably high flux of around 50 LMH. Contaminated groundwater is heated using low grade solar energy in an evacuated-type solar glass panel.

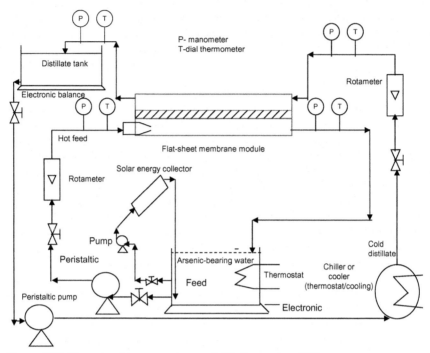

Figure 5.15 The solar-driven membrane distillation system [31].

Table 5.4 Typical Hydrophobic Membranes Useful in Membrane Distillation Module*

Membrane thickness	Material	Nominal pore size (μm)	Porosity (%)	Thickness (μm)	Active layer (μm)
MS3220	PTFE	0.22	80	150	60
MS3020	PTFE	0.22	80	175	60
MS7020	PP	0.22	35	160	160

*Manufactured by Membrane Solutions, Shanghai, China.

Hydrophobic, microporous flat–sheet membranes as presented in Table 5.4 are useful composite membranes where a composite membrane like MS 3220 is having a thin polytetrafluoroethylene (PTFE) active layer and a polypropylene (PP) support sublayer. MS3020 is another useful composite membrane of PTFE active membrane layer with polyethylene terephthalate (PET) supporting sublayer. MS7020 is a symmetric and isotropic membrane made of pure polypropylene that may also be used in a membrane distillation module.

The solar-driven membrane distillation system is illustrated in Figure 5.15. This solar-driven membrane distillation set-up consists of four major components: a direct contact membrane distillation module, a solar energy collector, and two thermostatic baths. The system works in two loops, namely the solar loop and the arsenic removal loop.

5.9.2 Solar heating loop

The solar loop consists of an evacuated glass tube type solar energy collector and a water-holding drum to heat and store arsenic-contaminated water. In this type, the solar collector may be made of double-layer borosilicate glass tubes (47 mm outer diameter, 37 mm inner diameter, 1.5 m length) evacuated (air is removed completely creating vacuum) for providing very good insulation. The outer wall of the inner tube is coated with selective absorbing material (black chrome) that absorbs solar radiation and transfers the heat to the arsenic-contaminated water that flows through the inner tube. Solar-heated water circulates between the storage tank and heating tubes continuously by natural convection and gravity (thermo-siphon process) until the sun sets. The brighter and stronger the radiation falling on the collector, the faster the circulation. Circulation of hot water between the feed tank and the collector storage tank is maintained by a centrifugal pump. The storage tank permits extended operation of the MD module even after sunset. Feed flow rate may be measured using a simple rotameter in the loop while a manometer registers the pressure in the collector storage tank (P_c). Temperatures of the feed inlet, outlet, and collector storage tank are monitored through the three attached dial thermometers.

5.9.3 Arsenic removal loop

The arsenic removal loop consists of two peristaltic pumps for circulation of the cold (distillate) and hot (feed) streams, two rotameters for measurement of these flow rates, and a flat-sheet direct-contact membrane distillation module operated in cross-flow mode. The new module is designed to hold a flat-sheet membrane. Hot feed water is pumped to the lower side of the membrane while cold stream water flows counter-currently over the upper surface of the membrane in the module. The effective membrane surface area in lab-made flat module design is 0.0120 m^2. The flat membrane module is horizontally oriented. Inlet and outlet pressures of the two streams are monitored through manometers. Dial thermometers indicate the module inlet and outlet temperatures of both streams. Thermostatic baths may be used for maintaining constant temperature of the streams during any study. However, this is not necessary during actual treatment of arsenic-contaminated water.

5.9.4 Operation of the membrane distillation module

Flat membrane sheets may be cut from a roll and sandwiched between two halves of a polycarbonate box. A finely perforated stainless steel plate supports the membrane in the module. The arsenic-contaminated water in the feed tank is heated with the aid of a solar heating loop and circulated to the arsenic removal loop by centrifugal pump at ground level. Feed side and permeate side temperatures may be controlled by controlling the ground level circulation pump and using thermostatic baths attached to the module.

Hot feed (arsenic-contaminated groundwater) is pumped to the lower side of the membrane (feed cell) while cold stream (pure distilled water) flowed counter-currently over the upper surface of the membrane (permeate cell) in the horizontally placed module. The thermocouples with a sensitivity of ± 273.1 K register the hot entrance T_{fb-in}, the cold entrance T_{pb-in}, the hot exit T_{fb-out}, and the cold exit T_{pb-out} of the membrane module. T_{fb} and T_{pb} are the average bulk feed and permeate temperatures estimated by taking the arithmetic mean of the temperatures at the inlet and outlet. A thermostatic bath may be used for maintaining constant desired temperatures of the cold distillate stream during experiment. Hydrophobic, microporous flat-sheet composite membranes (viz., MS3220 by Membrane Solutions, Shanghai, China) with a thin polytetrafluoroethylene (PTFE) active layer and polypropylene (PP) support sublayer and the second (MS7020) with a symmetric, isotropic membrane made of pure PP may be used. The PTFE (with 80% porosity) and PP (35% porosity) membranes have active layers of 60 μm and 160 μm, respectively, with the same nominal pore size (0.22 μm).

5.9.5 The operating conditions

5.9.5.1 Effect of feed temperature on flux

The effect of feed temperature on water flux is illustrated in Figure 5.16. The changes of viscosity and density of water over the given range of temperature variation are not so strong and as a result the Reynolds number variation with the feed temperature does not significantly change at a given flow rate.

Figure 5.16 illustrates that feed temperature has a remarkable influence on the permeate flux. For example, when the feed flow rate is 0.028 m/sec, an increase of temperature from 30 to 61 °C causes an increase in water vapor flux from all three different types of membranes. The fluxes of

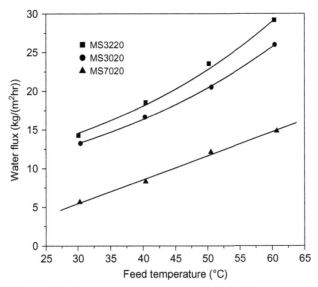

Figure 5.16 Effect of feed temperature on pure water flux [13].

MS3220 and MS3020 increase exponentially with temperature, whereas the flux through the MS7020 increases linearly with temperature. The exponential increase of MD flux with temperature for commercial PTFE membranes is due to the exponential increase of vapor pressure with temperature as per the Antoine equation as described in the modeling section for vapor pressure of water. The permeate flux exhibits a linear dependence on the feed temperature for PP membrane since mass transfer resistance of the membrane is mainly governed by membrane thickness. In the reported system [31], membrane thickness of PP is substantially greater (2.6 times) than PTFE. Thus the effect of the membrane thickness at higher temperature overshadows the effect of the temperature on water flux. The reasons for the higher flux of PTFE membranes other than PP are higher porosity of PTFE membranes and rougher surfaces (especially the surface of the support layer) of PTFE membranes that help mixing at the membrane interfaces.

5.9.5.2 Effect of feed velocity on flux and feed outlet temperature

Figure 5.17 shows the performance of the three membranes in terms of flux as a function of feed velocity.

Figure 5.17 illustrates how feed velocity influences flux and feed outlet temperature. Results show that flux increases with an increase of feed velocity. The reason is that an increase of feed velocity leads to an increase of

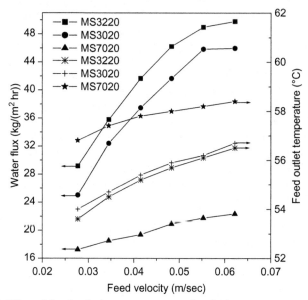

Figure 5.17 Effect of feed velocity on pure water flux [13].

Reynolds number, which in turn enhances the feed side boundary layer heat transfer coefficient by reducing the thickness of the temperature boundary layer. This leads to an increase of the feed side membrane surface temperature (T_{mf}) due to the reduction of feed side temperature polarization. This eventually results in a larger driving force for mass transfer through the membrane that supports the increased flux. Moreover, reduction of concentration polarization with an increase of feed velocity also helps to increase the flux, though the effect of temperature polarization is believed to be more pronounced than the effect of concentration polarization on flux.

Also, the composite PTFE membranes have higher flux values than the symmetric, isotropic PP membrane even though they have the same pore size. Probably higher membrane thickness of PP attributes to the lengthened path within the membrane for the vapor to pass through. Moreover, the PP membrane has lower porosity than the PTFE membrane. When feed flow velocity varied from 0.028 to 0.048 m/sec as per Figure 5.17, the flux increases from 29.16 to 46.24 kg/(m^2 hr) (i.e., about 59% for MS3220), 25 to 41.66 kg/(m^2 hr) (i.e., 66% for MS3020), and 17.27 to 20.93 kg/(m^2 hr) (22% for MS7020). But when the feed flow increases from 0.048 to 0.062 m/sec, the flux only increases from 46.24 to 49.8 kg/(m^2 hr) (8%) for MS 3220, 41.66 to 45.96 kg/(m^2 hr) (10%) for MS3020, and

20.93 to 22.36 kg/(m² hr) (6%) for MS7020. This suggests that there is an optimum feed flow rate for each membrane during the DCMD process. Therefore, for feed velocity values higher than 0.048 m/sec, no improvements in flux is observed due to absence of fluid dynamic control on the mass transfer. Thus, the effect of the feed flow rate to the membrane flux is not as significant as that of the feed temperature. On the other hand, the increase of feed velocity decreases the residence time of feed in the feed cell of the module and increases the feed outlet temperature, which increases the transmembrane vapor pressure difference due to the increase $\Delta T = T_{feed} - T_{distillate}$. This leads to a higher water vapor flux that condenses in the distillate side. As a result, the distillate outlet temperature also increases, which is reflected in Figure 5.17.

5.9.5.3 Effect of distillate velocity on flux and distillate outlet temperature

The effect of velocity of cold distillate flowing through the permeate side of the module on the water vapor flux and distillate outlet temperature is illustrated in Figure 5.18 for MS3220 and MS3020 membranes.

Figures 5.18 and 5.19 show the effect of distillate velocity on flux and distillate outlet temperature. The water vapor flux increases with an increase

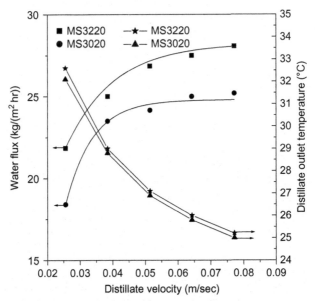

Figure 5.18 Effect of distillate velocity of pure water flux [13].

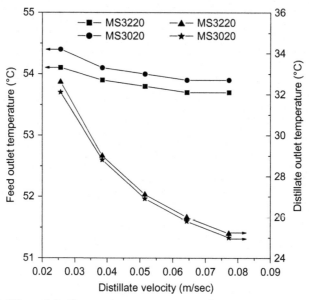

Figure 5.19 Effect of distillate outlet temperature on flux.

of distillate flow rate due to enhanced mixing in the flow channel and a decrease of the thickness in the temperature boundary layer. Therefore, the temperature at the permeate-membrane surface (T_{mp}) decreases, which reduces the temperature polarization on the cold distillate side. At a feed temperature of 60 °C, when the distillate flow rate varies from 0.026 to 0.052 m/sec, the water vapor flux increases from 21.85 to 26.86 kg/(m² hr) (23%) for MS3220 and 18.41 to 24.16 kg/(m² hr) (31%) for MS3020. On the other hand, at the same feed temperature (60 °C) and over the same variation of feed velocity (0.028–0.055 m/sec), the flux increases by 68% for MS3220 and 83% for MS3020. The results show that temperature polarization (TP) under the conditions of the cold distillate side is not as critical as it is for the feed side. Water vapor pressure at relatively low distillate temperature changes only slowly with temperature. Moreover, the effect on water vapor flux by concentration polarization with a varied distillate flow rate is considered to be zero due to absence of a concentration boundary layer in the distillate side. For a distillate flow rate higher than 0.052 m/sec, no improvement of flux is observed due to absence of fluid dynamic control on the mass transfer in the permeate cell.

Because of the decrease of the residence time of the distillate in the module with an increase of distillate velocity, the distillate outlet temperature

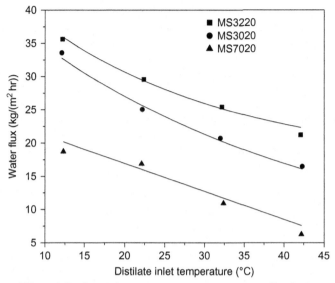

Figure 5.20 Effect of distillate inlet temperature on pure water flux [13].

decreases. As a result, the decreased distillate outlet temperature increases the vapor pressure-based driving force due to increased $\Delta T = T_{feed} - T_{distillate}$. With a decrease of distillate outlet temperature, water vapor flux increases. This leads to a decrease in the feed outlet temperature, which is indicated in Figure 5.19.

Figure 5.20 illustrates the performance of the three membranes at different distillate inlet temperatures across the membrane.

When the distillate inlet temperature increases from 12.1 to 42.3 °C keeping the feed inlet temperature constant at 60–61 °C, the distillate flux decreases from 35.62 to 21.2 kg/(m² hr) (40%) for MS3220, 33.56 to 16.47 kg/(m² hr) (51%) for MS3020, and 18.71 to 6.26 kg/(m² hr) (66%) for MS7020. This sharp decrease of flux following a drop of distillate inlet temperature by 30 °C is due to a decrease in the driving force for water vapor. Using two commercial PTFE membranes (MS3220 and MS3020), the exponential decrease of MD flux with an increase of distillate inlet temperature is observed. This is most probably due to the exponential decrease of vapor pressure with a rise in distillate temperature for commercial PTFE membranes; this may be attributed to a vapor pressure rise according to the Antoine equation. The most probable reason for a linear dependence of flux for PP membrane is that the mass transfer resistance is governed by membrane thickness.

5.9.5.2 Effect of arsenic concentration on flux

The effect of arsenic feed concentration on flux is depicted in Figure 5.21. At a feed temperature of 60–61 °C with a variation of arsenic concentration from 0 to 1200 ppb, the distillate flux decreased from 27.52 to 23.61 kg/(m² hr) (14%) for the MS3220 membrane, 25.23 to 21.85 kg/(m² hr) (13%) for MS3020, and 15.12 to 13.5 kg/(m² hr) (10%) for MS7020.

An average of 12% flux decline was observed in three membranes when arsenic concentration was increased from 0 to 1200 ppb [31]. This is attributed to the fact that the addition of the arsenic reduces the water activity of the feed solution since the water vapor pressure is the driving force of the MD process and it relates to the water activity. Thus, the reduction of flux following an increase of feed concentration is due to the decrease of the driving force, $p_{mf}-p_{mp}$. Moreover, an additional boundary layer, known as the concentration boundary layer, is formed adjacent to the membrane surface due to the presence of arsenic in feed. This concentration polarization further reduces the vapor pressure of water at the feed–membrane interface, thereby reducing the driving force for evaporation. However, the effect of the arsenic concentration on flux is relatively low because increased feed concentration only marginally decreases the vapor pressure in DCMD, as observed in several studies with other solutes.

Compared to RO, the effect of solute concentration of the feed water on flux is much lower in MD. With an increase of feed concentration in the RO process, the performance of the system reduces significantly due to a

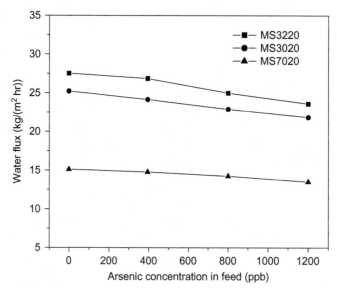

Figure 5.21 Effect of concentration of target solute in feed solution on pure water flux.

reduced driving force for mass transfer across the membrane resulting from increased salt passage, increased concentration polarization, scaling, and higher osmotic pressure.

5.9.5.3 Flux enhancement in new flash vaporization design

The high fluxes are achieved in the new flash vaporization design [13]. The membrane module is made of polycarbonate material having a high thermal insulation property that ensured minimization of heat exchange with the surroundings. Hot feed enters the wide feed side channel through a very narrow circular conduit (4 mm diameter) and undergoes flash vaporization upon exposure to low pressure in the wide channel. This ensures not only a high rate of evaporation but also promotes heat transfer through better mixing and minimization of temperature and concentration polarizations. The evaporation process is also facilitated by placing the feed cell at the bottom side of the module from which vapors can flow vertically upward through the microporous membrane to the cold distillate on the other side of the membrane. The distillate cell side channel has the same dimension as the feed channel and is designed for minimization of pressure drop in the flow channel and maximization of fluid mixing that reduced temperature polarization. Therefore, a large amount of vapor coming from the feed side were easily condensed in the cold distillate.

The simplicity of design, commercial availability of varieties of membranes, and abundance of solar energy of the most affected region of the world make the solar-driven membrane distillation system highly applicable to community-based purification of arsenic-contaminated groundwater, particularly in the South East Asian countries. The solar-driven system could yield reasonably high flux of around 50 kg/(m^2 hr). Membrane-wetting, which is considered a potential problem in membrane distillation, was almost negligible in the study despite prolonged use of the membrane. Since flux is one of the most significant aspects of successful implementation of SDMD, selection of the appropriate membrane through comparative study is very important.

5.10 MODELING AND SIMULATION OF A NEW FLASH VAPORIZATION MODULE FOR SCALE UP

5.10.1 Introduction

This section presents modeling and simulation of the solar-driven flash vaporization membrane distillation module [61]. The developed model is validated with rigorous experimental findings during membrane distillation of arsenic-contaminated groundwater. By incorporating flash vaporization dynamics the model turns out to be substantially different from the existing direct contact membrane distillation models, and succeeds in prediction of the system performance with a relative error of only 0.042 and a Willmott d-index of 0.997.

The flash-enhancing module is presented in Figure 5.22. The solar-driven flash vaporization membrane distillation (FVMD) set-up consists of two major components, a direct contact FVMD module and a solar heating loop made up of an evacuated glass panel.

The flat-sheet cross-flow membrane module made of polycarbonate sandwiched a flat membrane sheet of 10×10^{-4m} m^2 surface area between the feed cell and the permeate cell. The module here was designed with DCMD configuration with a special provision for flash vaporization. This solar-driven FVMD set-up works in two loops, a solar heating loop and a direct contact FVMD loop. An evacuated glass tube type solar energy collector is used to heat up the feed.

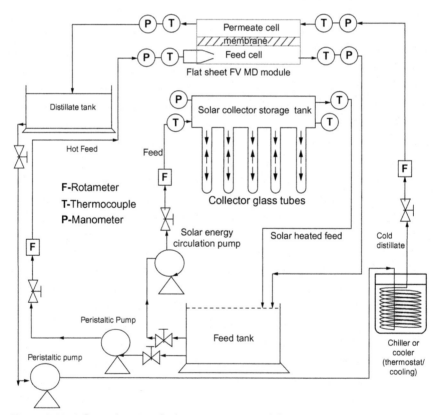

Figure 5.22 A flux-enhancing flash vaporization module [61].

Out of the four configurations in the MD processes, the DCMD is used most widely, in which hot liquid feed and cold liquid permeate are always in direct contact with the membrane surfaces. Mathematical models to predict the performance of MD modules of different configurations in terms of water flux, evaporation efficiency, and percentage rejection have been developed, mainly capturing the operating variables and membrane parameters and significant phenomena like temperature and concentration polarizations. These models, however, were developed in the context of desalination. Thus the model developed by Pal et al. [61] that deals with arsenic separation paves the way for optimization and scale up for new applications other than desalination.

Realistic mathematical models can be of great help in raising scale-up confidence. Thus, the present chapter deals with modeling and simulation of a flux-enhancing module.

5.10.2 Theoretical background of model development

The flux-enhancing module (Figure 5.22) is designated as FVMD. As the hot feed enters a relatively large feed cell through narrow inlet tubing, flash vaporization takes place due to a sudden pressure drop in the feed space. Due to this phenomenon, water vapor along with a very small quantity of air moves upward in the feed cell of the horizontally placed FVMD module. It then comes in contact with the feed side membrane surface whose temperature is below the dew point of the water vapor–air mixture, resulting from direct contact of cold distillate with the permeate side membrane surface. This leads to formation of a film of condensate on the feed side membrane surface along with an additional film of noncondensable gas (air) and vapor (water) of thickness δ_{fv} as illustrated in Figure 5.23.

The film condensate is considered to be the feed thermal boundary layer of thickness δ_{ft} as observed in the existing DCMD model, and the vapor–air film is situated between the bulk feed and the feed thermal boundary layer. In the permeate cell, formation of a permeate thermal boundary layer of thickness δ_{pt} takes place. Water vapor generated from the bulk feed by the flash vaporization process continuously mixes in the vapor–air film. Therefore, the temperature gradient across the vapor–air film is expected to be negligible. The difference in temperature between the two sides of the membrane gives rise to a difference in water vapor pressure between the feed and permeate sides, which drives water vapor through the membrane pores. Resistances against this water vapor transport arise from the presence of air in the membrane pores, presence of feed side and permeate

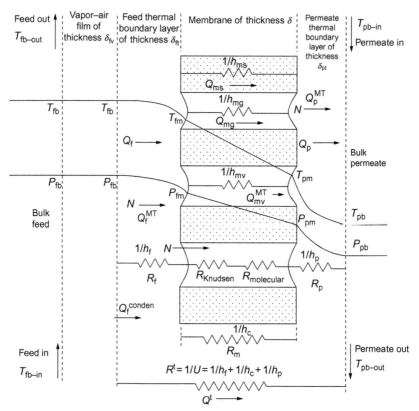

Figure 5.23 Mass and heat transfer resistances in membrane distillation module.

side thermal boundary layers, and the membrane. The overall separation and purification process here involves phase change and simultaneous mass and heat transfer. Thermal energy required for phase change (evaporation of water) at the feed–membrane interface is derived from the interface of the vapor–air film and feed thermal boundary layer. Water vapor in the vapor–air film condenses at this interface by releasing heat of condensation (Q_f^{conden}) that can be explained by Nusselt's condensing mechanism related to film-type condensation [62].

Vaporization at the feed–membrane surface is controlled by convective heat transfer Q_f and heat transfer due to mass transfer (Q_f^{MT}) as depicted in Figure 5.23. Water vapor condenses at the permeate–membrane surface after transport of vapor from the feed–membrane surface to the permeate-membrane surface through the membrane pores due to vapor pressure difference. This heat of condensation (Q_{mv}^{MT}) and membrane conductive heat (Q_{mc}) are transferred from the permeate-membrane interface to the bulk permeate across the permeate thermal boundary layer in the form of convective heat transfer (Q_p) and the heat transfer due to mass transfer (Q_p^{MT}). Simultaneous heat and mass transfer phenomena occurring in the MD process create temperature gradients in the two thermal boundary layers on two sides of the membrane. Old MD models consider these two thermal boundary layers. However, when the feed space is wide enough to cause flash vaporization, formation of an additional air–vapor film adjacent to the so-called feed thermal boundary layer takes place, facilitating continuous mixing of air and the water vapor. Eventually condensation of water vapor occurs following Nusselt's mechanism. Without consideration of this air–vapor film a mathematical model for the direct contact membrane distillation configuration system is unlikely to be capable of predicting the performance accurately for the designs where provision for flash vaporization is introduced for enhancing flux.

The model described here for the new flux enhancing membrane module with provision for flash vaporization considers the phenomenon of film condensation. The new design uses hydrophobic flat-sheet membranes to separate the flashing feed zone from the permeate chamber. An additional air–vapor film offers additional resistance to heat, and mass transfer is included in the present model development as described in the subsequent sections. Moving away from a desalination application, the new module is applied to a relatively new field of arsenic separation from contaminated groundwater and the experimental findings are used to validate the developed model.

5.10.3 Model development

The mathematical model for the flash vaporization module based on the transport mechanisms described in the previous section can now be developed with the assumption of applicability of the Antoine equation in vapor pressure build-up in view of the clean nature of groundwater, where on average, arsenic concentration (300–400 ppb) does not warrant failure of the Antoine equation. The other major reasonable assumption in view of the special design is obviously the existence of an additional air–vapor film on the feed side of the membrane in addition to the feed thermal boundary layer. Condensation of water vapor in the air–vapor film following Nusselt's condensation mechanism naturally gets incorporated in the model.

However, temperature gradient in the vapor–air film may be neglected because of the fact that a large amount of water vapor produced in the bulk feed by flash vaporization continuously mixes in the film. Vapor–air film in the feed side of the membrane module naturally maintains the same temperature of the bulk feed. This consideration in the modeling is justified by flash-induced turbulence and mixing. Finally vapor transport through the membrane may be considered to take place following the combined Knudsen and molecular diffusion mechanisms of the DGM. This is justified by the value of the Knudsen number ($K^n = \lambda/d$) that lies here between 0.01 and 1.0 at a typical membrane surface temperature of 333 K.

Transport of water vapor through the porous membrane of the DCMD system occurs by both diffusive as well as convective modes where trapped air acts against the transport. The Knudsen number ($K^n = \lambda/d$) in the present case lies between 0.01 and 1.0 at a typical membrane surface temperature of 333 K, and a combination of Knudsen and molecular diffusion mechanism in transport of water vapors may be assumed here. The air pressure (P_a being partial pressure, P_t being total pressure) needs to be considered for both membrane surfaces, and a logarithmic mean would be better representation of this resistance. The other resistance that encounters mass transfer of water vapor is the resistance of the membrane itself, which can be captured in the model through its structural properties like porosity (ε), tortuosity (τ), and thickness indicating path length (δ). Molecular weight (M) of water vapor, average membrane surface temperature(T_m), gas constant (R), Knudsen diffusion coefficient (D_k), and diffusivity of water vapor in air (D_{wa}) will be the other governing parameters in the diffusion and transport of water vapor through the membrane pores.

Similar to existing DCMD process, the total transport resistance (R^t) in the modified FVMD process is the sum of the resistances in the feed thermal boundary layer (R_f), resistance of the membrane (R_m), and resistance in permeate thermal boundary layer (R_p) (Figure 5.23). The resistance in the additional air–vapor film may be neglected because of continuous mixing of bulk feed side vapor in the film and thus the effect of temperature polarization may be considered negligible. The driving force for transport of the water vapor through the particular domain (feed or permeates thermal boundary layers or membrane) is the vapor pressure difference across that domain.

Therefore, feed thermal boundary layer resistance may be expressed as

$$R_f = \frac{p_{fb} - p_{fm}}{N} \tag{5.100}$$

Similarly, membrane resistance can be written as:

$$R_m = \frac{p_{fm} - p_{pm}}{N} \tag{5.101}$$

The combination of Knudsen and molecular diffusion resistances in series (Figure 5.23) is used to describe mass transfer across the membrane. Therefore, R_m can be expressed as the addition of the Knudsen and the molecular diffusion resistances:

$$R_m = R_{\text{Knudsen}} + R_{\text{molecular}} \tag{5.102}$$

where

$$R_{\text{Knudsen}} \text{ and } R_{\text{molecular}}$$

are written as:

$$R_{\text{Knudsen}} = \frac{\tau\delta}{\varepsilon}\frac{RT}{M}\left(\frac{1}{D_K}\right) \text{ and}$$

$$R_{\text{molecular}} = \frac{\tau\delta}{\varepsilon}\frac{RT}{M}\left(\frac{P_a}{P_t D_{\text{wa}}}\right) \tag{5.103}$$

Permeate boundary layer resistance can be evaluated by:

$$R_p = \frac{p_{\text{pm}} - p_{\text{pb}}}{N} \tag{5.104}$$

The total mass transfer resistance may now be expressed as:

$$R^t = R_f + R_m + R_p = \frac{p_{\text{fb}} - p_{\text{pb}}}{N} \tag{5.105}$$

Membrane structural properties and physical properties of the permeating substance may be combined into the membrane distillation coefficient for the Knudsen-molecular diffusion mechanism [63]:

$$\text{MDC} = \frac{\varepsilon}{\tau\delta}\frac{M}{RT_m}\left(\frac{1}{D_K} + \frac{P_a}{P_t D_{\text{wa}}}\right)^{-1}\text{ln} \tag{5.106}$$

where the subindex ln indicates the logarithmic mean of $\{1/D_K + p_a/(P_t D_{\text{wa}})\}^{-1}$ on both the membrane surfaces, where different values of air pressure (P_a) exist.

The driving force in water vapor transport will be the difference between feed–membrane surface water vapor pressure (P_{fm}) and permeate–membrane surface water vapor pressure (P_{pm}). These vapor pressures will in turn depend on interface temperatures T_{fm} and T_{pm} at the feed side and permeate side, respectively.

Thus the initial flux equation for water vapor may be expressed as:

$$N = \frac{\varepsilon}{\tau\delta}\frac{M}{RT_m}\left(\frac{1}{D_K} + \frac{P_a}{P_t D_{\text{wa}}}\right)^{-1}_{\text{ln}}\left(p_{\text{fm}} - p_{\text{pm}}\right) \tag{5.107}$$

This FVMD model despite resemblance with the DCMD model is expected to predict flux in a substantially different way than the in the way the classical DCMD model does owing to different approaches adopted in computing related film coefficients, interface temperatures, and hence the vapor pressures on two sides of the membrane. Such differences arise due to consideration of the additional air–vapor film on the feed side. Heat transfer in FVMD may be described in four distinct zones (Figure 5.23): vapor–air film, feed thermal boundary layer, membrane, and permeate thermal boundary layer. Heat of condensation, Q_f^{conden}, liberates at the interface of vapor–air film and feed thermal boundary layer due to condensation of vapor of this film obeying the Nusselt's condensing mechanism. In the second step, the total heat transfer across the feed thermal boundary layer Q_f^t is the summation of the convective heat transfer in the feed thermal boundary layer, Q_f, and the heat transferred due to mass transfer across the feed boundary layer, Q_f^{MT} (Figure 5.23). Q_f and Q_f^{MT} may be written as:

$$Q_f = h_f(T_{fb} - T_{fm}) \tag{5.108}$$

$$Q_f^{MT} = NH_{L,f}\left\{\frac{T_{fb} + T_{fm}}{2}\right\} \tag{5.109}$$

The total heat Q_p^t is removed from the permeate-membrane surface by the bulk permeate in the form of the convective heat transfer in the permeate boundary layer, Q_p, and the heat transferred due to mass transfer across the permeate boundary layer, Q_p^{MT} (Figure 5.23), where

$$Q_p = h_p\left(T_{pm} - T_{pb}\right) \tag{5.110}$$

and

$$Q_p^{MT} = NH_{L,p}\left\{\frac{T_{pb} + T_{pm}}{2}\right\} \tag{5.111}$$

In Eqs. (5.110) and (5.111) [12–15], $H_{L,f}$ is the enthalpy of the feed that is evaluated at the mean temperatures of the feed and membrane surface, $(T_{fb} + T_{fm})/2$; $H_{L,p}$ is the enthalpy of the permeate that is evaluated at the mean temperatures of feed and membrane surface, $(T_{pb} + T_{pm})/2$; N is the mass flux; h_f is the feed thermal boundary layer heat transfer coefficient; and h_p is the permeate thermal boundary layer heat transfer coefficient.

The heat, Q_m, transfers across the membrane in the form of latent heat because of water vapor migration through the membrane pores, Q_{mv}^{MT}, and membrane conductive heat transfer, Q_{mc}. Q_{mc} is actually the summation of conductive heat transfer through the membrane solid polymer, Q_{ms}, and air-vapor trapped inside the membrane pores, Q_{mg}. Temperature distribution across the membrane is a nonlinear temperature distribution and the flow is assumed to be nonisenthalpic flow. In that case, the heat flux equation may be expressed as

$$Q_m = Q_{mc} + Q_{mv}^{MT} = -k_m \frac{dT}{dX} + NH_v\{T\} \tag{5.112}$$

where k_m is the thermal conductivity of the polymeric membrane, and can be expressed as

$$k_m = (1 - \varepsilon)k_s + k_g\varepsilon \tag{5.113}$$

where ε is the porosity, k_s and k_g are the thermal conductivities of the hydrophobic membrane solid polymer and air-water vapor trapped inside the membrane pores, respectively.

At steady-state conditions, the overall heat transfer flux Q^t through the FVMD system may be expressed as

$$Q_f + Q_f^{MT} = Q_m = Q_p + Q_p^{MT} = Q^t \tag{5.114}$$

By introducing the additional air–vapor film, we try to find out heat transfer effects of mass transfer across the thermal boundary layers and contributions of various components of heat transfer to the total heat flux. From the proposed FVMD model equations, initially heat transfer due to mass transfer in the feed thermal boundary layer (Q_f^{MT}), in the permeate thermal boundary layer (Q_p^{MT}), and across the membrane (Q_{mv}^{MT}) may be computed as shown in Table 5.5. Contributions of these components to the total heat flux (Q^t) can then be found out.

Table 5.5 Mass Transfer Contribution to the Overall Heat Flux for PTFE and PP Membranes at Different Temperatures and at Feed and Distillate Velocities of 0.062 m/s and Distillate Temperature of 294 K

PTFE membrane (MS3220)				PP membrane (MS7020)			
Feed temperature (K)	$\frac{Q_f^{MT}}{Q^t}\%$	$\frac{Q_p^{MT}}{Q^t}\%$	$\frac{Q_{mv}^{MT}}{Q^t}\%$	Feed temperature (K)	$\frac{Q_f^{MT}}{Q^t}\%$	$\frac{Q_p^{MT}}{Q^t}\%$	$\frac{Q_{mv}^{MT}}{Q^t}\%$
303.1	1.40	1.14	29.25	303.4	0.72	0.56	15.11
313.4	2.56	1.77	39.60	313.6	1.34	0.85	20.95
323.2	4.02	2.55	50.27	323.9	2.29	1.26	28.32
333.3	5.83	3.52	61.25	333.5	3.50	1.77	36.52

Source: Reference [61].

From these computations, it is found that $\%Q_f^{MT}$, $\%Q_p^{MT}$, and $\%Q_{mv}^{MT}$ increase with feed temperature. But $\%Q_p^{MT} < \%Q_f^{MT} \ll \%Q_{mv}^{MT}$, and the difference of any two quantities increases with increase of feed temperature. The increase of Q_f^{MT} is more significant than Q_p^{MT} due to the effect of mass flux higher at a higher feed temperature. At 333 K, the maximum percentage contribution was 5.83% for Q_f^{MT} and 3.52% for Q_p^{MT}, and the minimum percentage contribution was at around 303 K at 15.11% for Q_{mv}^{MT}. Therefore, contributions of Q_f^{MT} and Q_p^{MT} to total heat flux can be neglected from the proposed FVMD model. Contributions of both Q_f^{MT} and Q_p^{MT} for a PTFE membrane were more than that of a similar pore size PP membrane since a higher membrane thickness of the PP membrane attributes to the lengthened path within the membrane for the vapor to pass through. Moreover, the PP membrane has lower porosity than the PTFE membrane. The Table 5.5 also shows that the percentage of Q_{mv}^{MT} for PTFE is the major component at high temperatures (above 323 K), while the conduction heat transfer (Q_{mc}) through the membrane is the major component for PP at both high and low temperatures.

Thus from these initial calculations and with the assumptions of linear temperature distribution across the membrane and isenthalpic flow of vapor, Q_m may be written as

$$Q_m = Q_{mc} + Q_{mv}^{MT} = h_m\left(T_{fm} - T_{pm}\right) + N\Delta H_v \qquad (5.115)$$

where h_m is the heat transfer coefficient of the polymeric membrane:

$$h_m = \{(1-\varepsilon)k_s + k_g\varepsilon\}/\delta \qquad (5.116)$$

Here, the enthalpy of vapor, $H_v\{T\}$, is nearly equal to the latent heat of vaporization, ΔH_v.

At steady-state conditions neglecting Q_f^{MT} and Q_p^{MT} according to the previous discussion, the overall heat transfer flux Q in the modified FVMD model may be obtained from Eq. (5.115):

$$Q_f = Q_m = Q_p = Q$$
$$\text{i.e., } h_f\left(T_{fb} - T_{fm}\right) = h_m\left(T_{fm} - T_{pm}\right) + N\Delta H_v = h_p\left(T_{pm} - T_{pb}\right) \qquad (5.117)$$

From Eq. (5.117), T_{fm} and T_{pm} for the modified FVMD model are represented as $T_{fm\ mod}$ and $T_{pm\ mod}$ and they can be evaluated by the following equations:

$$T_{fm\,mod} = \frac{h_m\left\{T_{pb} + \left(h_f/h_p\right)T_{fb}\right\} + h_f T_{fb} - N\Delta H_v}{h_m + h_f\left(1 + h_m/h_p\right)} \qquad (5.118)$$

$$T_{\text{pmmod}} = \frac{h_m\left\{T_{fb} + \left(h_p/h_f\right)T_{pb}\right\} + h_p T_{pb} - N\Delta H_v}{h_m + h_p\left(1 + h_m/h_f\right)} \tag{5.119}$$

Using the model equations of the flash vaporization membrane distillation as described in this section, the values of R (resistance), U (overall heat transfer coefficient), TPC (temperature polarization coefficient), VPC (vapor pressure polarization coefficient), and evaporation efficiency may be computed easily.

U for the modified FVMD process may be expressed as

$$U = \left[\frac{1}{h_f} + \frac{1}{h_m + \left(N\Delta H_v\right)/\left(T_{\text{fmmod}} - T_{\text{pmmod}}\right)} + \frac{1}{h_p}\right]^{-1} \tag{5.120}$$

The temperature polarization coefficient (TPC), which represents loss of thermal driving force due to thermal boundary layer resistances for FVMD model, may be expressed as

$$\text{TPC} = \frac{T_{\text{fmmod}} - T_{\text{pmmod}}}{T_{fb} - T_{pb}} \tag{5.121}$$

The temperature and concentration polarization result in the reduction of the effective driving force and this reduction in driving force, which is expressed in the MD literature as vapor pressure polarization coefficient (VPC), can now be shown as

$$\text{VPC} = \frac{p_{fm} - p_{pm}}{p_{fb} - p_{pb}} \tag{5.122}$$

One of the efficiency parameters of the MD process is EE, defined as the ratio of the heat transferred because of vapor migration through the membrane pores to the total heat transferred through the membrane. Following the standard definition in literature, EE for the modified FVMD model may be expressed as

$$EE = \frac{Q_{mv}^{MT}}{Q_{mv}^{MT} + Q_{mc}} = \frac{\Delta H_v N}{N\Delta H_v + h_m\left(T_{\text{fmmod}} - T_{\text{pmmod}}\right)} \tag{5.123}$$

5.10.4 Solutions to the model equations

Model equations are solved following an iterative method and using the physico-chemical parameters obtained from empirical relations as described in the next section. Experimental investigations as described in the previous section were carried out for validation of the model. A statistical correlation analysis was done to examine how model predictions corroborate with the experimental findings. Statistical correlation analysis was carried out to estimate the error and assess the fitness of the model through use of root mean square error (RMSE), relative error (RE), and the Willmott d-index (d). RMSE, RE, and d were estimated by using the following equations [64]:

$$RSME = \sqrt{\frac{\sum_{i=1}^{n}(P_i - E_i)^2}{n}} \tag{5.124}$$

$$RE = \frac{RSME}{\overline{E}} \tag{5.125}$$

$$d = 1 - \frac{\sum_{i=1}^{n}(P_i - E_i)^2}{\sum_{i=1}^{n}\{|(P_i - \overline{E})| + |(E_i - \overline{E})|\}^2} \tag{5.126}$$

5.10.5 Determination of physico-chemical parameters
5.10.5.1 Diffusion coefficient
D_K is the Knudsen–diffusion coefficient of water vapor, which is expressed as [14]

$$D_K = \frac{2}{3}r\left(\frac{8RT}{\pi M}\right)^{1/2} \tag{5.127}$$

where r is the pore radius of the membrane.

The value of $P_t D_{wa}$ (Pa m^2/sec) was calculated from the following expression [65]:

$$P_t D_{wa} = 4.46 \times 10^{-6} T^{2.334} \tag{5.128}$$

where T is the absolute temperature.

5.10.5.2 Water vapor pressure
Pure water interfacial vapor pressures p_{fm} and p_{pm} were calculated using the Antoine equation:

$$p_{im}^o = \exp\left(23.238 - \frac{3841}{T_{im} - 45}\right), i = f, p \tag{5.129}$$

where T_{im} is the corresponding interfacial temperature in Kelvin.

5.10.5.3 Enthalpy of liquid and vapor

To calculate water enthalpy, $H_L\{T\}$, at temperature T of interest (at feed or permeate side), the following equation was used from the enthalpy data fitting for water given in the thermodynamic properties table [65] in the temperature range 273–373 K.

$$H_L\{T\} = (4.1863\,T - 1143.4)10^3 \tag{5.130}$$

To determine the enthalpy of vapor, $H_v\{T\}$, at temperature T, the following equation was also determined from the enthalpy data fitting for saturated water vapor taken from a thermodynamic properties table [65] in the temperature range 273–373 K.

$$H_v\{T\} = (1.7535\,T + 2024.3)10^3 \tag{5.131}$$

5.10.5.4 Film transfer coefficients in the feed side (h_f) and permeate side (h_p)

In the DCMD literature, feed side film transfer coefficient (h_f) is calculated using empirical equations. In our modified membrane distillation module we have provision for flash vaporization and the model considers an additional air–vapor film as shown in Figure 5.23. Assume that water vapor condensation in this zone occurs following Nusselt's condensation mechanism. The Nusselt's equation [65,66] below was used in deriving the film coefficient h_f for the present new FVMD model.

$$h_f = 0.725 \left(\frac{g\Delta H_{vfb}\rho_f^2 k_f^3}{\mu_f w (T_{fb} - T_{fm})} \right)^{0.25} \tag{5.132}$$

where ρ_f, k_f, and μ_f are the density, thermal conductivity, and viscosity of feed in the feed thermal boundary layer, respectively. ΔH_{vfb} is the latent heat of vaporization and was calculated for bulk feed temperature T_{fb}, acceleration due to gravity g and the width of feed channel w.

Permeate side film transfer coefficient h_p for the laminar flow ($Re < 2,100$) was evaluated by the empirical equation used in the existing DCMD model since the permeate cell in the FVMD module is designed in such a way that the fluid dynamics in the permeate cell in the present study are assumed to be similar to that observed in the existing DCMD processes; thus the following equation holds for the permeate side:

$$Nu_p = 3.66 + \frac{0.104 Re_p \, Pr_p \left(d_{hp}/L\right)}{1 + 0.0106 \left[Re_p \, Pr_p \left(d_{hp}/L\right)\right]^{0.8}} \tag{5.133}$$

where Nu_p, Re_p, and Pr_p are the Nusselt, Reynolds, and Prandtl numbers at the permeate side; d_{hp} permeate side hydraulic diameter; and L permeate channel length.

$$Nu_p = \frac{h_p d_{hp}}{k_p}, Re_p = \frac{v_p d_{hp} \rho_p}{\mu_p}, Pr_p = \frac{c_{pp} \mu_p}{k_p} \tag{5.134}$$

where h_p, k_p, ρ_p, μ_p, and c_{pp} are the heat transfer coefficient, the thermal conductivity, the density, the viscosity, and the heat capacity of the permeate thermal boundary layer, respectively. v_p is the permeate velocity.

In the existing DCMD literature, h_f is computed using empirical equations as

$$Nu_f = 3.66 + \frac{0.104 Re_f \, Pr_f \left(d_{hf}/L\right)}{1 + 0.0106 [Re_f \, Pr_f \left(d_{hf}/L\right)]^{0.8}} \tag{5.135}$$

where Nu_f, Re_f, and Pr_f are the Nusselt, Reynolds, and Prandtl numbers at the feed side; d_{hf} feed side hydraulic diameter; and L feed side channel length.

$$Nu_f = \frac{h_f d_{hf}}{k_f}, Re_f = \frac{v_f d_{hf} \rho_f}{\mu_f}, Pr_f = \frac{c_{pf} \mu_f}{k_f} \tag{5.136}$$

where h_f, k_f, ρ_f, μ_f, and c_{pf} are the heat transfer coefficient, the thermal conductivity, the density, the viscosity, and the heat capacity of the feed thermal boundary layer, respectively. v_f is the feed velocity.

5.10.5.5 Membrane-feed (T_{fm}) and membrane-permeate (T_{pm}) interfacial temperatures

Membrane interfacial temperatures for FVMD model (i.e., T_{fm} and T_{pm}) are computed with the help of Eqs. (5.137 and 5.138).

$$T_{fm} = \frac{cj - bl}{aj - bd} \tag{5.137}$$

$$T_{pm} = \frac{al - dc}{aj - bd} \tag{5.138}$$

where

$$a = -h_f - \frac{1753.5 N e^A}{e^A - 1}, \quad b = \frac{1753.5 N}{e^A - 1},$$

$$c = -h_f T_{fb} - N H_{L,f} \left\{ \frac{T_{fb} + T_{fm}}{2} \right\} + 2024.3 \times 10^3 N$$

$$d = -\frac{1753.5 N e^A}{e^A - 1}, \quad j = h_p + \frac{1753.5 N}{e^A - 1},$$

$$l = h_p T_{pb} - N H_{L,p} \left\{ \frac{T_{pb} + T_{pm}}{2} \right\} + 2024.3 \times 10^3 N$$

where

$$A = \frac{1753.5 N \delta}{k_m}$$

The interface temperatures T_{fm} and T_{pm} and flux (N) may be computed iteratively or using standard numerical methods. The numerical values of T_{fm}, T_{pm}, and N can be predicted for the given experimental conditions T_{fb}, T_{pb}, and v, when the characteristics parameters of the membrane r, ε, τ, k_m, and δ are known. In the iterative procedure, initial values of T_{fm} and T_{pm} are assumed and N is calculated from the flux Eq. (5.139):

$$N = \text{MDC} . \Delta p \tag{5.139}$$

where N is water vapor flux through the membrane pores and MDC is the membrane distillation coefficient. Δp is the driving force of mass transfer being the vapor pressure difference between the feed thermal layer–membrane interface and permeate thermal layer–membrane interface. Then Eqs. (5.137) and (5.138) are used to calculate T_{fm} and T_{pm} after determining the film transfer coefficients (h_f and h_p) from Eqs. (5.132) through (5.136), and liquid enthalpies ($H_{L,f}\{T\}$ and $H_{L,p}\{T\}$) from Eq. (5.130). The obtained values are compared with the initially assumed values. If they are different, a second iteration is done assuming that the values for T_{fm} and T_{pm} are the ones obtained from the previous iteration. The calculation is repeated until the assumed values for T_{fm} and T_{pm} agree with the calculated ones to the desired degree.

5.10.5.6 Membrane interfacial temperatures for modified FVMD model

The equations of the modified FVMD model are used to evaluate the numerical values of $T_{fm\ mod}$, $T_{pm\ mod}$, and flux (N) by applying an iterative method described in the previous section. Determination of h_f value is first done considering the existing DCMD feed cell (where no vapor–air film included is considered in the model for the FVMD process) instead of Nusselt's equation. The iteration method is also used in the existing DCMD model equations to calculate the values of $T_{\text{fm mod}}$, $T_{\text{pm mod}}$, and flux (N) following the same procedures as applied in the FVMD model. $T_{\text{fm mod}}$ and $T_{\text{pm mod}}$ are considered in place of T_{fm} and T_{pm} in the equations related to the modified FVMD model. The VPC value at a certain feed temperature of feed is evaluated by the following method algorithm.

1. Calculation of $T_{\text{fm mod}}$ and $T_{\text{pm mod}}$ from Eqs. (5.118) and (5.119) for the modified FVMD model.
2. Calculation of vapor pressures p_{fm}, p_{pm}, p_{fb}, and p_{pb} at temperatures $T_{\text{fm mod}}$, $T_{\text{pm mod}}$, T_{fb}, and T_{pb}, respectively, from the Antoine Eq. (5.129).
3. Calculation of VPC from Eq. (5.122).

5.10.6 Simulation

5.10.6.1 Effect of feed temperature on flux and vapor pressure

The mathematical model of flash vaporization membrane distillation as developed by Pal et al. [61] for separation of arsenic from contaminated groundwater has been found to very successfully predict performance of a real system as illustrated in Figure 5.24. Figure 5.25 illustrates how closely

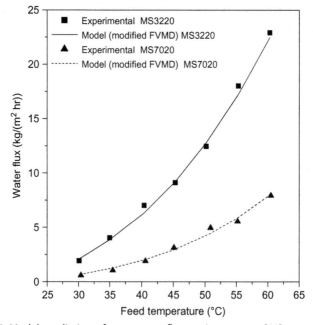

Figure 5.24 Model prediction of pure water flux against system [61].

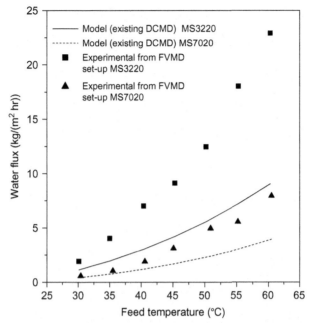

Figure 5.25 Model prediction versus system performance (water flux) [61].

the modified FVMD model agrees with real system performance in terms of flux during change of feed temperature, whereas Figure 5.26 shows the closeness of an existing DCMD model with the same flash vaporization system performance. Comparing these two figures, it becomes obvious that a modified FVMD model is much more realistic than the existing DCMD model in predicting a new flash vaporization membrane distillation system for arsenic separation.

From the feed temperature versus water flux graphs, it transpires that mass flux increases with feed temperature almost exponentially as expected in accordance with the Antoine equation. The increase of temperature from 303 to 334 K causes an increase of pure water vapor flux but the increase is much more for the MS3220 membrane than the MS7020 membrane. The higher flux (two to three times) for MS3220 than for MS7020 can be traced to higher porosity, lower membrane thickness, and rougher surfaces of MS3220, especially the surface of the support layer that helps mixing at the membrane interfaces.

Table 5.6 presents statistical correlation analysis of simulated flux with real system flux for both the classical DCMD model and the modified FVMD model. Comparison reflects very good performance [23] of the

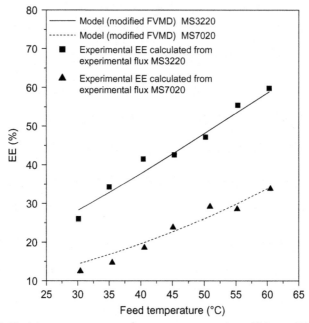

Figure 5.26 Model versus system performance: evaporation efficiency [61].

Table 5.6 Comparative Model Performance in Terms of Flux through Statistical Correlation Analysis

	MS3220			MS7020			
Model	RMSE	RE	Willmott d-index	RMSE	RE	Willmott d-index	Remarks
FVMD	0.458	0.042	0.998	0.260	0.072	0.997	Model performance is very good as $R < 0.1$ and d-index > 0.95
Existing DCMD model	7.579	0.703	0.648	2.120	0.592	0.745	Very poor performance and not acceptable

Source: Ref. [61].

modified FVMD model (Willmott d-index being >0.95 and relative error being <0.1) against poor fitness of the classical DCMD model.

Figure 5.26 depicts how the model predicted EE agrees with system performance [61].

In case of both membranes, EE increases with an increase of feed temperature because flux increases exponentially with temperature. With an increase of feed temperature from 303 to 334 K, system EE increases by 129% for MS3220 and 171% for MS7020. Also the value of EE increases with an increase in the porosity from MS7020 to MS3220 because of its dependence on process vapor flux, N. At a feed temperature of 333 K, EE of the system is almost 60% for MS3220 (PTFE membrane) in the FVMD module against 42% in the existing DCMD process for the same membrane. In statistical correlation analysis, relative error (RE) and Willmott d-index for MS3220 are found to be 0.042 and 0.992, respectively, while the corresponding values for membrane MS7020 were 0.043 and 0.984, indicating a very good fitness of the model (Willmott d-index being >0.95 and relative error being <0.1).

5.10.6.2 Effect of feed temperature on temperature polarization coefficient and vapor pressure polarization coefficient

Temperature polarization coefficient values (TPC) computed after evaluation of $T_{\text{fm mod}}$ and $T_{\text{pm mod}}$ from the modified FVMD model are presented in Figure 5.27. This figure shows TPC for the FVMD system decreases with an increase of feed temperature.

This is due to the exponential rise of the vapor pressure that increases the permeate flux substantially as the temperature rises. These larger mass fluxes mean that a larger amount of heat transfer occurs through the feed and permeate side liquid phases, increasing the temperature gradient in the boundary liquid layers, and so the temperature polarization (TP). Due to the high heat transfer coefficient in the feed side compared to that of the permeate side, TP in the permeate side ($T_{\text{pm mod}} - T_{\text{pb}}$) is much higher than that in the feed side ($T_{\text{fb}} - T_{\text{fm mod}}$).

In the existing DCMD literature, the highest values of TPC reported for the pure water is within a range of 0.4 (high fluxes) to 0.7 (low fluxes). The low value of TPC in the flash vaporization module results from higher permeate side temperature polarization (TP) as illustrated in Figure 5.28. Figure 5.28 also shows that the TPC for MS7020 is more or less double that for MS3220. This is attributed to much lower flux obtained in MS7020 when compared with MS3220. Permeate side TP can be reduced by

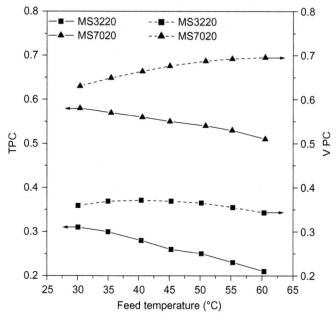

Figure 5.27 Simulated TPC and VPC [61].

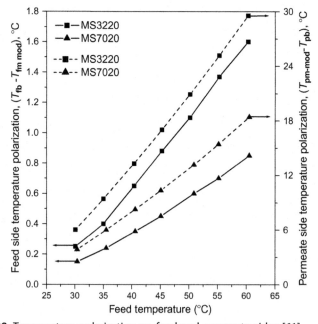

Figure 5.28 Temperature polarization on feed and permeate sides [61].

improving the permeate side design of flow passage or membrane arrangement, or by applying turbulence promoters like mesh spacers.

The VPC and different vapor pressure polarizations ($p_{fb} - p_{fm}$, $p_{pb} - p_{pm}$, $p_{fm} - p_{pm}$) of the two types of membranes at different feed temperatures have been presented in Figures 5.27 and 5.28, respectively. The driving force (p_{fb}-p_{pb}) is the same for PTFE and PP in the temperature interval (303–334 K). Figure 5.27 depicts the simulated values of VPC for PP membrane increase with an increase of feed temperature, whereas the VPC value for PTFE increases slowly up to 314 K and then decreases when the temperature further increases. This is because the rate of increase of (p_{fm}-p_{pm}) is more than the rate of increase of (p_{fb}-p_{pb}) with an increase of feed temperature. This is true for the PP membrane for the entire studied temperature range. It is also observed that VPC values for the PP membrane is always higher than that for the PTFE membrane. As per Figure 5.28, the value of T_{pm-mod} for PTFE is always greater than those values for the PP membrane due to higher flux for the PTFE membrane. As a result larger values of (p_{fm}-p_{pm}) for PP compared to PTFE leads to higher values of VPC for the PP membrane.

5.10.6.3 Effect of feed temperature on heat transfer coefficients and transport resistances

Figure 5.29 shows heat transfer coefficients calculated from a model, and illustrates how h_f decreases with an increase of feed temperature, which is the opposite trend documented in various existing MD references [12]. This trend might be explained with the help of Nusselt's equation number, in which h_f is inversely related to $(T_{fb\ mod} - T_{fm\ mod})^{1/4}$ and proportionally related to $(\Delta H_{vfb})^{1/4}$. Figure 5.29 shows $(T_{fb} - T_{fm\ mod})$ increases with an increase of feed temperature. On the other hand, ΔH_{vfb} decreases with increase of feed temperature as per general trend. The temperature dependence of the other physical properties might increase h_f with an increase of temperature, but this effect is much less over the effect of $(T_{fb} - T_{fm-mod})$ and ΔH_{vfb}. On the other hand, h_p increases with an increase of feed temperature following the same trend as noticed in various existing DCMD studies. This is due to the permeate cell of the FVMD module having been designed similar to the permeate cell of the existing DCMD process. Figure 5.29 also depicts the effect of temperature, which is stronger on h_f than h_p. With a feed temperature increase from 303 to 334 K in the case of MS3220 membrane, h_f decreased by 24.6%, while h_p and U increased by 1.5% and 19.13%, respectively. This is also true for the PP membrane. The large variation of h_f

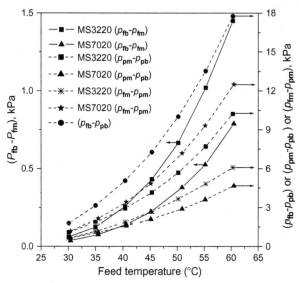

Figure 5.29 Effect of feed temperature on transport resistances [61].

and h_p is due to the fact that the variation of the temperature on the feed side is more important than on the permeate side. The considerably very high value of h_f (an order of magnitude of 10^4 W/m^2 K) in the laminar flow in the FVMD module is more or less the same order of the h_f value in the turbulent flow in the existing DCMD process (10, 12). On the other hand, h_p value of the FVMD process is considerably low (less than 1000 W/m^2 K) in the laminar flow. This is because the flash vaporization concept is adopted in the feed cell whereas the permeate cell is designed according to the existing DCMD process.

5.11 TECHNO-ECONOMIC FEASIBILITY OF USE OF SOLAR MEMBRANE DISTILLATION

Though membrane distillation theoretically offers the possibility of removal of arsenic from contaminated water by 100%, it suffers from the problem of low flux. However, development of the new flash vaporization membrane module by Pal and Manna [31] has been successful in more than doubling this flux. Emergence of tailor-made membranes at a cheaper price, development of flux enhancing modules, and the worldwide emphasis of the use of sunlight as an alternate renewable energy resource have brightened the possibility of the use of solar-driven membrane distillation in vast arsenic-affected areas of the world, particularly the South East Asian countries blessed with abundant sunlight almost throughout the year. Using solar energy for heating water to only 60–70 °C and for running small pumps for household or community-based water treatment units are no longer a techno-economic challenge, but are real, distinct possibilities.

At a feed temperature of 344 K, distillate temperature of 277.4 K, and optimum distillate velocity of 0.123 m/sec, water vapor flux of around 52 kg/(m^2 hr) could be achieved. This marks a substantial improvement over the reported earlier results. Membrane distillation offers pure water free from not only arsenic but also from all sorts of pathogens in the backdrop of huge problem of water-borne diseases in the arsenic-affected areas. Thus benefits are manifold.

Membrane distillation (MD) has been gaining importance in the recent years, in purification of water and purification of thermo-labile substances compared to the conventional processes due to involvement of low operating temperature and pressure, potential of high degree of separation, negligible membrane fouling, and emergence of new membranes. Enhanced process safety, flexibility in using plastic materials for fabrication of the membrane module, and the scope for using low-grade energy like solar or geothermal energy are some other major advantages of MD.

The developed mathematical model could predict the performance of the newly developed flash vaporization module with a high degree of accuracy, raising the confidence of scale up. It transpires that solar-driven FVMD can be a potential technology in separation of arsenic from contaminated groundwater. Techno-economic feasibility is very clear and obvious.

NOMENCLATURE

c_p	Fluid specific heat (J/kg/K)
d	Willmott d-index
d_h	Hydraulic diameter of membrane module (m)
D_k	Knudsen-diffusion coefficient of water vapor (m^2/sec)
D_{wa}	Diffusion coefficient of water vapor in air (m^2/sec)
DCMD	Direct contact membrane distillation
EE	Evaporation efficiency (%)
FVMD	Flash vaporization membrane distillation
g	Acceleration due to gravity (m/sec^2)
H	Enthalpy (J/kg)
$H_L\{T\}$	Liquid enthalpy at temperature T (J/kg)
$H_v\{T\}$	Vapor enthalpy at temperature T (J/kg)
h	Heat transfer coefficient (W/(m^2 K))

ΔH_{vfb}	Heat of vaporization of water at bulk feed temperature (J/kg)
k	Thermal conductivity (W/(m K))
K^n	Knudsen number
L	Chamber length (m)
ln	Logarithmic mean
M	Molecular weight of water (kg/mol)
MD	Membrane distillation
MDM	Membrane distillation coefficient
N	Water flux (kg/(m^2 sec))
Nu	Nusselt number
p	Partial pressure (Pa)
P_a	Partial pressure of air (Pa)
P_t	Total pressure (Pa)
p_{im}^o	Interfacial vapor pressure of pure water (Pa)
Pr	Prandtl number
Q	Heat flux (W/m^2)
Q_f^{conden}	Heat transfer to the feed boundary layer from vapor–air film due to condensation (W/m^2)
Q_{mc}	Conduction heat transfer rate through membrane (W/m^2)
r	Pore radius (m)
R	Gas constant (J/mol/K)
Re	Reynolds number
R_f	Resistance of feed boundary layer (Pa m^2 h/kg)
R_m	Resistance of the membrane (Pa m^2 h/kg)
R_p	Resistance of permeate boundary layer (Pa m^2 h/kg)
R^t	Total resistance (Pa m^2 h/kg)
RE	Relative error
RMSE	Root mean square error
T	Temperature (K)
$T_{fm\ mod}$	Modified temperature at the feed–membrane surface
$T_{pm\ mod}$	Modified temperature at the permeate–membrane surface
T_m	Average membrane surface absolute temperature inside the pores
TP	Temperature polarization
TPC	Temperature polarization coefficient
U	Overall heat transfer coefficient (W/(m^2 K))
v	Fluid velocity (m/sec)
VPC	Vapor pressure polarization coefficient
w	Width of the feed channel (m)

Greek letters

δ	Membrane thickness (m)
δ_{ft}	Thermal feed boundary layer thickness (m)
δ_{pt}	Thermal permeate boundary layer thickness (m)
δ_{fv}	Vapor-air film thickness (m)

ε Porosity
μ Viscosity (Pa s)
ρ Density (kg/m^3)
τ Tortuosity

Subscripts

f   Feed
i   Feed or permeate
fb   At the bulk phase of the feed side
fm   At the membrane surface of the feed side
L   Liquid phase
m   Membrane
mg   In the membrane pores related to air vapor
ms   In the membrane solid material
mv   In the membrane pores related to water vapor
p   Permeate
pb   At the bulk phase of the permeate side
pm   At the membrane surface of the permeate side
s   Solid phase

Superscripts

MT   Heat transfer due to mass transfer
t   Total

REFERENCES

[1] Godino P, Peña L, Mengua JI. Membrane distillation: theory and experiments. J Membr Sci 1996;121:83–93.
[2] Gryta M. Influence of polypropylene membrane surface porosity on the performance of membrane distillation process. J Membr Sci 2007;287:67–78.
[3] Hogan PA, Canning RP, Johnson RA. Osmotic distillation: new options. Chem Eng Progr 1998;34:49–68.
[4] Bailey JL, McCune RF (Polaroid). Microporous vinylidene fluoride polymer and process of making same. US Patent 3642668; 1972.
[5] Grandine II JD (Millipore). Process of making a porous membrane material from Polyvinylidene fluoride, and products. US Patent 4203848; 1983.
[6] Tomaszewska M. Preparation and properties of flat-sheet membranes from poly(vinylidene fluoride) for membrane distillation. Desalination 1996;104:1–11.
[7] Kinzer KE, Shipman GH, Yang YF, Lloyd DR, Tseng HS. Thermally-induced phase separation membranes: structure-formation relationships. In: Proceedings of the international congress on membranes and membrane processes, Tokyo; 1987. p. 344–5.
[8] Bierenbaum HS, Isaacson RB, Lantos PR. Breathable medical dressing. US Patent 3426754; 1969.

[9] Soehngen JW, Ostrander K (Celanese). Solvent stretch process for preparing a microporous film. US Patent 4257997; 1981.

[10] Fisher HM, Leone DE (Hoechst Celanese). Microporous membranes having increased pore densities and process for making the same. EP Patent 0342026 B1; 1989.

[11] Gostoli C, Sarti GC. Thermally driven mass transport through porous membranes. In: Sedlacek B, Kahovec J, editors. Synthetic polymeric membranes. Berlin: Walter de Gruyter and Co; 1987b. p. 515–29.

[12] Lawson KW, Lloyd DR. Membrane distillation. J Membr Sci 1997;124:1–25.

[13] Lei Z, Chen B, Ding Z. Special distillation process. The Netherlands: Elsevier Science; 2005.

[14] El-Burawi MS, Khayet M, Ma R, Li Z, Zhang X. Application of vacuum membrane distillation for ammonia removal. J Membr Sci 2007;301:200–9.

[15] Mason EA, Malinauskas AP. Gas transport in porous media: the dusty gas model. New York: Elsevier; 1983.

[16] Bonyadi S, Chung TS. Flux enhancement in membrane distillation by fabrication of dual layer hydrophilic–hydrophobic hollow fiber membranes. J Membr Sci 2007;306:134–46.

[17] Lagana F, Barbieri G, Drioli E. Direct contact membrane distillation: modelling and concentration experiments. J Membr Sci 2000;166:1–11.

[18] Schofield RW, Hogan PA, Fane AG, Fell CJD. Developments in membrane distillation. Desalination 1987;62:728.

[19] Banat FA, Simandl J. Theoretical and experimental study in membrane distillation. Desalination 1994;95:39–52.

[20] Sarti GC, Gostoli C, Matulli S. Low energy cost desalination process using hydrophobic membranes. Desalination 1985;56:277–86.

[21] Martinez-Diez L, Florido-Diaz FJ, Hernandez A, Pradanos P. Estimation of vapor transfer coefficient of hydrophobic porous membranes for applications in membrane distillation. Sep Purif Technol 2003;33:45–55.

[22] Gryta M. Long-term performance of membrane distillation process. J Membr Sci 2005;265:153–9.

[23] Karakulski K, Gryta M. Water demineralization by NF/MD integrated processes. Desalination 2005;177:109–19.

[24] Jonsson AS, Wimmerstedt R, Harrysson AC. Membrane distillation: a theoretical study of evaporation through microporous membranes. Desalination 1985;56:237–49.

[25] Khayet M, Godino P, Mengual JI. Nature of flow on sweeping gas membrane distillation. J Membr Sci 2000;170:243–55.

[26] Basini G, Angelo D, Gobbi M, Sarti GC, Gostoli C. A desalination process through sweeping gas membrane distillation. Desalination 1987;64:245–57.

[27] Banat FA, Simandl J. Removal of benzene traces from contaminated water by vacuum membrane distillation. Chem Eng Sci 1996;51(8):1257–65.

[28] Qu D, Wang J, Hou D, Luan Z, Fan B, Zhao C. Experimental study of arsenic removal by direct contact membrane distillation. J Hazard Mater 2009;163:874–9.

[29] Macedonio F, Drioli E. Pressure-driven membrane operations and membrane distillation technology integration for water purification. Desalination 2008;223:396–409.

[30] Islam AM. Membrane distillation process for pure water and removal of arsenic [MSc thesis]. Chalmers University of Technology; 2005.

[31] Pal P, Manna AK. Removal of arsenic from contaminated groundwater by solar-driven membrane distillation using three different commercial membranes. Water Res 2010;44:5750–60.

[32] Schofield RW, Fane AG, Fell CDJ. Heat and mass transfer in membrane distillation. J Membr Sci 1987;33:299–306.

[33] Curcio E, Drioli E. Membrane distillation and related operations. A review. Separ Purif Rev 2005;34:35.

[34] Schofield RW, Fane AG, Fell CJD. Gas and vapour transport though microporous membrane. II. Membrane distillation. J Membr Sci 1990;53:173–85.

[35] Srisurichan S, Jiraratananon R, Fane AG. Mass transfer mechanisms and transport resistances in direct contact membrane distillation process. J Membr Sci 2006;277:186–94.

[36] Qtaishat M, Matsuura T, Kruczek B, Khayet M. Heat and mass transfer analysis in direct contact membrane distillation. Desalination 2008;219:272–92.

[37] Gryta M, Tomaszewska M, Morawski AW. Membrane distillation with laminar flow. Sep Purif Technol 1997;11:93–101.

[38] Martinez L, Florido-Diaz FG. Theoretical and experimental studies on desalination using membrane distillation. Desalination 2001;139:373–9.

[39] Gryta M, Tomaszewska M. Heat transport in membrane distillation process. J Membr Sci 1998;144:211–22.

[40] Kimura S, Nakao S. Transport phenomena in membrane distillation. J Membr Sci 1987;33:285–98.

[41] Phattaranawik J, Jiraratananon R, Fane AG. Heat transport and membrane distillation coefficients in direct contact membrane distillation. J Membr Sci 2003;212:177–93.

[42] Smolders CA, Franken ACM. Terminology for membrane distillation. Desalination 1989;72:249–62.

[43] Criscuoli A, Carnevale MC, Drioli E. Evaluation of energy requirements in membrane distillation. Chem Eng Process: Process Intesif 2007;47:1043–50.

[44] Martinez L, Rodriguez-Maroto JM. Characterization of membrane distillation modules and analysis of mass flux enhancement by channel spacers. J Membr Sci 2006;274:123–37.

[45] Sherwood TK, Pigford RL, Wilke CR. Mass transfer. Chemical eng. series, McGraw Hill; 1975.

[46] Schofield RW, Fane AG, Fell CJD. Gas vapor transport through microporous membranes. I. Knudsen-Poiseuille transition. J Membr Sci 1990;53:159–71.

[47] Martinez L, Rodriguez-Maroto JM. On transport resistances in direct contact membrane distillation. J Membr Sci 2007;295:28–39.

[48] Khayet M, Godino MP, Mengual JI. Study of asymmetric polarization in direct contact membrane distillation. Sep Sci Technol 2004;39(1):125–47.

[49] Viegas RMC, Rodriguez M, Luque S, Alvarez JR, Coelhoso IM, Crespo JPSG. Mass transfer correlations in membrane extraction: analysis of Wilson-plot methodology. J Membr Sci 1998;145:129–42.

[50] Bodalo A, Gomez J, Gomez E, Leon G, Tejera M. Ammonia removal from aqueous solutions by reverse osmosis using cellose acetate membranes. Desalination 2005;184:149–55.

[51] Khayet M, Matsuura T. Preparation and characterization of Polyvinylidene fluoride membranes for membrane distillation. Ind Eng Chem Res 2001;40:5710–8.

[52] El-Bourawi MS, Khayet M, Maa R, Ding Z, Li Z, Zhang X. Application of vacuum membrane distillation for ammonia removal. J Membr Sci 2007;301:200–9.

[53] Alklaibi AM, Lior N. Heat and mass transfer resistance analysis of membrane distillation. J Membr Sci 2006;282:362–9.

[54] Bandini S, Sarti GC. Heat and mass transfer resistances in vacuum membrane distillation per drop. AIChE J 1999;45(7):1422–33.

[55] Mason EA, Malinauskas AP. Gas transport in porous media: the dusty gas model. New York: Elsevier; 1983.

[56] Albert RA, Silbey RJ. Physical chemistry. 2nd ed. New York: Wiley; 1997.

[57] Cussler EL. Diffusion, mass transfer in fluid system. 2nd ed. New York: Cambridge University Press; 1997.

[58] Ding Z, Ma R, Fane AG. A new model for mass transfer in direct contact membrane distillation. Desalination 2002;151:217–27.

[59] Bird RB, Stewart WE, Lightfoot EN. Transport phenomena. New York: Wiley; 1960.

[60] Schofield RW, Fane AG, Fell CJD. Gas and vapor transport through microporous membranes, I. Knudsen-Poiseuille transition. J Membr Sci 1990;53:159–71.

[61] Pal P, Manna AK, Linnanen L. Arsenic removal by solar-driven membrane distillation: modeling and experimental investigation with a new flash vaporization module. Water Environ Res 2013;85:63–76.

[62] Ding Z, Liu L, El-Bourawi MS, Ma R. Analysis of a solar-powered membrane distillation system. Desalination 2005;172:27–40.

[63] Schofield RW, Fane AG, Fell CJD. Gas and vapor transport though microporous membrane. II. Membrane distillation. J Membr Sci 1990;53:173–85.

[64] Stockle CO, Kjelgaard J, Bellocchi G. Evaluation of estimated weather data for calculating Penman-Monteith reference crop evatranspiration. Irrig Sci 2004;23:39–46.

[65] Perry RH. Perry's chemical engineer's handbook. 6th ed. New York: McGraw–Hill; 1984.

[66] Mason EA, Malinauskas AP. Gas transport in porous media: the dusty gas model. New York: Elsevier; 1983.

CHAPTER 6

Disposal of Concentrated Arsenic Rejects

Contents

6.1 INTRODUCTION

Treatment of contaminated water for production of potable water eventually leads to generation of concentrated rejects that need to be contained to avoid recycling contaminants back into the aquifers. Under increasing pressure of evergrowing human population, volumes of arsenic rejects from the treatment plants increase commensurate with the increase in size of the treatment plant. From physicochemical treatment plants, volumes of arsenic rejects in the form of sludge become quite considerable though membrane-based treatment plants produce relatively small volumes of concentrated arsenic rejects. However, in both cases, the rejects need to be treated further for safe disposal. Without carrying out the task of this final step, no treatment plant can be considered sustainable.

Scientific concentrate disposal is an integral part of a successful and viable treatment plant. With the progress of time, increased environmental awareness and more stringent regulations are likely to impose more restrictions on disposal options. Thus, a modern treatment plant must attach adequate

Groundwater Arsenic Remediation
http://dx.doi.org/10.1016/B978-0-12-801281-9.00006-0

importance to safe disposal of concentrated arsenic rejects. If arsenic–bearing solid sludge passes the TCLP (Toxicity Characterstics Leaching Procedure) test and CWET (California Wet Extraction Test) successfully, it is classified as nonhazardous waste and it can be disposed safely through landfill. Such sludge, if used in filling low-lying areas and in filling the rural roadsides, will not lead to any environmental problem. This stabilized sludge may also be used in manufacturing red bricks. Arsenic-contaminated sludge may also be used as a substitute for clay material used in manufacturing special quality bricks (ornamental bricks). Arsenic- and iron-bearing sludge can be used in enhancing compressive strength of clay bricks.

Due to limited options for disposal of highly concentrated arsenic rejects, many arsenic removal plants generally dump the rejects into the environment, potentially risking recontamination of underground aquifers through the natural percolation process. A possible solution to this disposal problem may be traced to stabilization of arsenic in some solid matrix. Coagulation and coprecipitation with other minerals eventually binding arsenic in an insoluble form with such coprecipitators can lead to a viable solution. Such coagulation–precipitation involving salts of iron, aluminum, and calcium have been reported in the literature. pH, molar ratio, and mineral combinations very much influence stabilization process. Thus, the stabilization parameters need to be properly optimized, eliminating the mutual interaction effects. Ferric arsenate is not thermodynamically stable at higher ranges of pH (>8) because it forms a solid layer of ferric hydroxide on the ferric arsenate precipitate and thus reduces the precipitation efficiency. Attempts have been made to improve the effectiveness of stabilization of different arsenic compounds using some solidification binders like Portland cement. But the excess lime present in cement provides an alkaline surface on the cement matrix, enhancing the instability of the ferric arsenate and causing the high cost of stabilization. The disposal volume also increases sharply, inviting further problems in transportation. Thus, the problem of instability of such solid precipitates still remains to be solved.

Stabilization/solidification (S/S) is a process normally used as a prelandfill waste treatment technology for the safe disposal of waste. The process involves mixing the waste in the form of sludge, liquid or solid, into a cementations binder system. S/S is most suitable for treating wastes that are predominantly inorganic, as these are considered more compatible with the cementations binders used. The aim is to encapsulate and incorporate the waste into the binder system and produce a monolithic solid with improved structural integrity that exhibits long-term stability and minimal leaching. S/S technologies inhibit leaching of hazardous components by reducing waste/leachate contact and by forming a stable pH environment in which many heavy metals of environmental concern remain insoluble.

Arsenic compounds (arsenic trioxide, arsenic pentoxide, sodium arsenite, and sodium arsenate) can be solidified by using different solidification binders such as fly ash, polymeric materials, Portland cement, combined Portland cement and iron (II) sulfate, and combined Portland cement and lime (CaO). Arsenic is also stabilized by simultaneous use of ferric and calcium salts that help in turning it into a nonhazardous, insoluble solid compound. Inorganic geopolymers can encapsulate arsenic in a similar manner to cement binders but this can be used as nonstabilized material. Synthesis of geopolymers is possible using waste materials rich in Al_2O_3 or SiO_2, which are mixed in specific ratios with alkali metal hydroxides and the major components of activated alumina removal units.

6.2 CLASSIFICATION OF ARSENIC-BEARING WASTES GENERATED IN DIFFERENT TREATMENT PLANTS

Three major categories of arsenic-bearing wastes are normally generated in three broad classes of treatment schemes. These are (1) solid aluminium or iron-based precipitate containing arsenic, (2) solid sorbate material with deposited arsenic on the surface and in the pores, and (3) concentrated aqueous solution from membrane-based plants with a high concentration of dissolved arsenic.

The class (1) arsenic waste is generated in huge volumes in physicochemical treatment plants where chemical coagulation–precipitation is the mechanism by which arsenic is transferred from an arsenic-contaminated water stream to the iron- or aluminium-based precipitates. The class (2) waste is the exhausted adsorbent material of adsorption-based plants. Often such adsorbent material is subjected to thermal regeneration for recovery of adsorption capacity of the adsorbate for its reuse. In the thermal regeneration process, part of the arsenic may escape into the atmosphere. The concentrated rejects from membrane-based plants are relatively less voluminous and after removal of water, the leftover solid volume will be very small, and subsequently can be bound to suitable materials through chemical coagulation–precipitation. Thus, all removal technologies of arsenic generate an arsenic-rich waste, either as a true precipitate, a sorbent, or concentrated rejects to which the arsenic is bound. Table 6.1 presents the details of typical arsenic-bound residuals generated from the treatment of arsenic-bearing groundwater.

Table 6.1 Characteristics of Arsenic-Bearing Residuals from the Treatment Plants of Arsenic-Contaminated Groundwater [1]

Technology involved	Residual type	Residual volume (gal/MG)	Arsenic concentration (mg/L)	Solids produced (lb./MG)	Arsenic in solids (mg/Kg)
Conventional coagulation	Sludge	4,300	9.3	180	1,850
Chemical softening	Sludge	9,600	4.2	2,000	165
Ion exchange	Liquid	4,000	10	23.4	14,250
Coagulation– microfiltration	Sludge	52,600	0.8	11	3,000
NF or RO	Liquid	664,000	0.1	NA	NA

Source: Reference [1].

Options for safe disposal of arsenic-bearing sludge are limited. In many cases, untreated arsenic contaminated waste is simply dumped or buried, leaving the potential danger of leaching arsenic back into the environment. Although studies have shown that arsenic leaching from sludge is not always significant, the validity of the leaching tests used is often questionable as they do not simulate true field conditions. Recycling is not a viable option due to limited uses and markets for As. High-temperature thermal treatment processes such as incineration volatilize arsenic-containing compounds, producing hazardous aerosols or an arsenic-containing sludge from cleaning emissions. Thus, the only sustainable management option for arsenic-containing waste is to convert the arsenic into the least mobile or stabilized form and isolate the stabilized material from the environment using a solidification/encapsulation process. Arsenic cannot be destroyed; it can only be converted to an insoluble compound or less harmful solid form by chemically combining it with other elements or binders such as iron, cement, polymer, and so on.

6.3 ARSENIC WASTE DISPOSAL METHODS

To make arsenic-contaminated water treatment plants sustainable, arsenic waste disposal must be effective and reliable for long-term operation. Disposal of arsenic-based hazardous waste requires effective management, so wherever the waste is generated there is a need to establish a well-defined protocol for its safe disposal. In general, there are four options available for dealing with arsenic waste streams:
• Dilution and dispersion through landfill
• Volatilization of arsenic by mixing with livestock waste

- Encapsulation or stabilization of the arsenic residuals
- Reuse as construction material by suitable binding with cement-type material

The first two options are often adopted by mining industries because of the possibility for combining numerous waste streams with such arsenic wastes, thus diluting the hazardous contaminants. This may help pass any regulatory limit but does not represent a scientific approach to solve the real problem. This may be a legislative solution but not a technical solution. Serious health problems, including enhanced risks of skin cancers and various internal carcinomas, could happen with long-term exposure of high concentrations of arsenic waste [2].

Currently, the most attractive option for dealing with arsenic waste lies in stabilizing or encapsulating the contaminated waste, usually through coagulation and precipitation techniques and disposing treated wastes in secure landfills. A detailed arsenic stabilization method including response surface optimization has been discussed in Chapter 4.

6.3.1 Dilution and dispersion through landfill

The disposal of arsenic-bound sludge or waste at a landfill is a possible way to dispose of arsenic rejects. But such sludge should be freed from water. Furthermore, the solid residuals must not have toxic characteristics as defined by the TCLP test or CWET [1].

In the absence of clear guidelines for safe disposal, arsenic-bound solid wastes are often disposed of in the open environment. Small sand-covered brick-lined pits are commonly used [3] for dumping sorptive filter media and regenerative wastes of an adsorption-based water treatment plant. Air drying can be adopted for a high water-containing arsenic sludge generated by coprecipitation and adsorption units. Sealed pits could be used to prevent flooding and leaching out of the arsenic into the surrounding soil and groundwater. TCLP and column leaching tests of arsenic-bound waste materials are also conducted.

In disposal scenarios of arsenic sludge, examination of the leaching behavior of waste is a crucial step in the environmental assessment. In many cases, landfill waste does not pass the TCLP test because the bonding of arsenic species with the sorptive or coagulate media is not stable. This is a major limitation of this method. Sorption of arsenic by ferric hydroxide gets reduced in presence of organic matter (at $pH > 8$) [4]. Organic acids are formed by anaerobic reduction of organic species, whereas buffering maintains pH due to dissociation of large volumes of dissolved CO_2. Redox zonation of landfill forms anaerobic, transition, and aerobic zones [5]. Ferric iron (Fe-III) is reduced to ferrous iron (Fe-II) within the anaerobic zone, and under this reducing condition of the environment, at pH 3–8, arsenate is reduced to the more mobile arsenite [6].

6.3.2 Volatilization of arsenic by mixing with livestock waste

Mixing of arsenic sludge with livestock waste is another disposal method. Some studies [7,8] have suggested that cow dung can be used as livestock waste to eliminate arsenic in volatile form, where the microorganisms present in cow dung reduce soluble arsenic species to gaseous arsine (AsH_3) and release it into the atmosphere. The method that eventually transfers arsenic from water to air in a more toxic form (arsine gas) in large volumes cannot be considered a sustainable one.

6.3.3 Encapsulation or stabilization of the arsenic residuals

Two widely used techniques for the encapsulation/stabilization of arsenic wastes are solidification via cementation and solidification via matrix compound formation. These two techniques are generally used as a prelandfill waste treatment technology for safe disposal of hazardous waste. The processes involve mixing the arsenic-bearing solid or liquid waste with a cementing binder system or an inorganic/organic compound that aims to encapsulate and incorporate arsenic waste into the binder system. This produces a monolithic solid with improved structural integrity that exhibits long-term stability and minimal leaching. Entrapping arsenic in a solid matrix through coagulation–precipitation prevents its leaching out through chemical interaction as a stable pH environment is formed in which arsenic compounds remain largely insoluble.

6.3.4 Encapsulation or stabilization via cementation

One of the most popular techniques due to its low cost is basically an encapsulation or stabilization technology used to transform potentially hazardous liquid or solid wastes into less hazardous or nonhazardous solids before disposal in a landfill, thus preventing the waste from entering the environment. The resultant alkaline pH renders arsenic insoluble.

Solid wastes arising from activated alumina-based adsorption units can stabilize As(III) with Portland cement, fly ash, and polymeric materials such as polymethylmethacrylate and polystyrene [9]. Such stabilization results from precipitation of calcium arsenite and calcite, which seal the pores in the solidified waste. However, the addition of polymers as cylindrical beads (1 mm diameter and 2.5 mm length) increases the possibility of arsenic leaching. Different solidified waste matrices can be prepared by mixing a variety of binder materials with arsenic-bearing waste in different ratios. Different such possible compositions are shown in Table 6.2. Water is mixed in a definite proportion with the solid matrix to make a slurry of these binders and arsenic-rich waste materials. To evaluate leaching potential, TCLP and semidynamic leach tests are conducted. A low degree of leaching out of arsenic from such a solid matrix is attributed largely to a formation of calcite and calcium arsenite. Minimum leaching out of arsenic is observed from a matrix having a mixture of activated alumina, cement, fly ash, and calcium hydroxide where formation of calcite takes place.

Three different types of binders [10] such as Portland cement, a mixture of Portland cement and iron sulfate, and a mixture of Portland cement and lime (CaO) were used to investigate the effectiveness of stabilization of four different arsenic compounds (arsenic trioxide, arsenic pentoxide, sodium arsenite, and sodium arsenate). Sequential batch leaching tests were conducted as per using the Australian Bottle Leaching Procedure guideline [11] at a neutral pH of extraction fluid with a test duration of 18 hr. Increases in Ca leachate concentrations were associated with decreasing arsenic concentrations and from the study result they concluded that Ca influenced the leaching of cement-immobilized arsenic, with formulations containing higher Ca:As mole ratios generally resulting in lower arsenic leaching. The efficiency of the stabilization process also depended on the treated arsenic compound, and it was found that arsenate compounds have the lowest mobility. The results showed that arsenic leaching ranged from 510 mg/L (arsenic acid) to 1.7 mg/L (arsenate).

Table 6.2 Composition Ratio of Different Solid Waste Samples

Matrices composition	Weight ratio
Activated alumina + cement	3:1
Activated alumina + cement + fly ash	3:1:0.5
Activated alumina + cement + fly ash + calcium hydroxide	3:1:0.5:0.5
Activated alumina + cement + fly ash + polystyrene	3:1:0.5:0.5
Activated alumina + cement + polystyrene	3:1:0. 5
Activated alumina + cement + polymethylmethacrylate	3:1:0.5

Source: Reference [9].

Arsenic ions have also been stabilized by [12] cement. The formed arsenic–calcium silicate hydrated matrix, however, is much more complicated in structure than calcium arsenate compounds. The matrix compound is formed due to adsorption or coprecipitation of arsenic ions with calcium and silicate compounds present in the cement. But at low pH (around 4), the matrix loses its stability and releases arsenic.

As_2O_3 has also been used [13] as an activator in the Vetrocoke technology and the arsenic-bearing waste material arising from the carbon dioxide scrubbing is precipitated as calcium and ferric arsenates or in arsenite form, and is encapsulated in a cement matrix. In this study [13], 100 g of wet sludge and an oxidation agent of 30% H_2O_2 solution, 40% ferric sulfate solution, and powder form of calcium oxide are mixed together precipitating arsenic in an insoluble compound form. $Na_2HAsO_4 \cdot 7H_2O$ could be immobilized [14] for long periods in a Portland cement matrix.

A mixture of supplementary cementitious materials such as slags and ashes with Portland cement has been found to produce a stable matrix [15]. A TCLP test of arsenite and arsenate after 28 days cure compared with 3 years cure shows no appreciable change, for a number of additives. The combination of Portland cement and fly ash however, shows increasing leaching out (by TCLP) with time and respeciation during curing. These results highlight the importance of long-term testing to identify specific combinations of stabilization binders and wastes that may undergo respeciation and consequent changes in leaching. Longer curing periods enhance oxidation of As(III) and its further immobilization in cement matrix [16].

The cementation method using Portland cement, lime, and fly ash has been found to effectively stabilize arsenic [17] for its safe disposal. To produce low permeability materials, the water:cement ratio needs to be kept as low as possible since water plays a significant role in initiating cement hydration reactions. Excess water may segregate initially well-mixed slurries, causing bleed water containing soluble waste components. Thus, the water:cement ratio plays an important role while dealing with arsenic sludge obtained from coagulant and coprecipitative units. Further studies have demonstrated that other components present in wastes may accelerate or retard set, even at low concentrations [18].

Portland cement is found to be the best binder, suggesting that unlike arsenate-bearing wastes, arsenite-saturated wastes may be effectively bound using Portland cement without the addition of lime. But the major limitation of these techniques is that the cementing binders used often constitute a major cost component in the process and therefore, the minimum amount that enables the stabilization waste to pass short-term regulatory test requirements for compressive strength and leaching normally is used. The waste may adversely affect the setting reactions of the binder, and this together with the highly complex chemistry of solid wastes, makes the prediction of long-term performance and durability difficult [18]. Additionally, excess lime present in cement provides an alkaline surface on the cement matrix, enhancing instability of the ferric arsenate and involving high cost of stabilization. The disposal volume also increases sharply, inviting further problems in transportation. Thus, the problem of instability of such solid precipitates still remains to be solved.

6.3.5 Encapsulation or stabilization via matrix compound formation

Encapsulation or stabilization of arsenic wastes by matrix compound formation using slags, polymers, and inorganic salts (such as iron or calcium) is another safe disposal option for arsenic waste.

6.3.6 Encapsulation or stabilization of arsenic residuals by industrial slags

Slag matrices may be used [19] to stabilize the arsenic from copper smelter flue dust. The stabilization that takes place in the presence of lime and at a low air roasting temperature converts arsenic oxide contained in the flue dust to calcium arsenate and arsenite. The calcium arsenate and arsenite compounds are dissolved in a molten iron silicate slag matrix. The leaching of arsenic from such slag matrix is found to be less than the maximum EP Toxicity Test limit of 5 ppm.

Arsenic can also be stabilized by lead–zinc blast furnace slags to some extent by using a calcium arsenite-containing waste [20]. To dissolve the arsenic into the slag, the arsenic-bearing waste needs to be mixed with the slag and heated to 1300–1400 °C. However, leaching out potential of As(III) remains high and unacceptable.

Dutre and Vandecasteele [21] developed a solidification process for stabilization of arsenic. The process requires proper mixing of blast furnace slags (having 5 M of HCl, zinc and iron concentration of 60 g/L, and lead concentration of 150 g/L), slaked lime, cement, and water. A duration of over 48 hours is required for setting the solid matrix. A 1:1:1 ratio of arsenic waste, cement, and lime yields the best result. Silicic acid and the silicate compounds of the binder materials cause polymerization in a slow process that needs a long setting time.

Recently, an encapsulation process has been developed [22] to stabilize arsenic from fly ash waste of a metallurgical industry that contains 23–47% by weight of arsenic. Typically, 8 g of lime, 6 g of cement, and 20 mL of water per 10 g of waste material are used as an optimum stabilization condition to achieve the highest arsenic stabilization efficiency. The solidification process is capable of reducing the leachate concentration from 5 g/L to approximately 5 mg/L. The stabilization of arsenic is attributed to formation of the $CaHAsO_3$ compound in the leachate, in the presence of $Ca(OH)_2$.

6.3.7 Encapsulation or stabilization of arsenic residuals by polymer compounds

Polymeric matrix may be used to encapsulate solid adsorbent-based arsenic residuals. Arsenic containing granular ferric oxy/hydroxide and ferric hydroxide amended alumina residuals are encapsulated in a polymeric matrix using an aqueous-based manufacturing process [23]. Arsenic loading in three different adsorbent is shown in Table 6.3. The polymer used as a stabilizing agent is a mixture of styrene butadiene and epoxy resin. The polymeric waste forms produced are capable of containing more than 60 wt% of sorbent on dry basis. Comparison of the waste form developed here with conventional cement matrices containing the same residuals show that the polymeric matrices are capable of encapsulating appreciably more material, but the stability of such a matrix is less than that of a cement-based matrix.

Inorganic geopolymers have also been found to encapsulate arsenic in a similar manner to cement binders [24]. Synthesis of geopolymers is possible using waste materials rich in Al_2O_3 or SiO_2, which are mixed in specific ratios with alkali metal hydroxides; the major component of activated alumina removal units commonly used in Bangladesh and India is Al_2O_3. It would seem likely that, given the nonmetallic character of arsenic, its role as a network former in a geopolymer system would be somewhat limited.

Table 6.3 Water Content, Arsenic Loading, and Equilibrium Arsenic Concentration for Sorbents Used in the Investigation

Sorbent used	Water content (wt%)	As loading/g of sorbent (mg/g)	As equilibrium concentration (μg/L)
GFH	45.6	3.8	35
E-33	60.4	8.0	48
AAFS	46.5	2.7	21

Source: Reference [23].

Little published information exists to confirm or dispute this, but it seems probable that arsenic would be immobilized by physical occlusion, rather than incorporation into the geopolymer structure.

6.3.8 Encapsulation or stabilization of arsenic residuals by inorganic salts

A possible solution to this disposal problem may also be traced in stabilization of arsenic with other minerals in some solid matrix. Coagulation and coprecipitation with other minerals eventually binding arsenic in an insoluble form with such coprecipitators can lead to a viable solution. In such a solution, coagulation–precipitation involving salts of iron, aluminum, and calcium have been reported in the literature. In such studies, pH, molar ratio, and mineral combinations have been found to influence the stabilization process.

Of the successful solidification or stabilization formulations, the use of iron appears to be the most preferred option. Ferric salts ($FeCl_3$ and $Fe_2(SO_4)_3$) can be used as coagulants, leading to the precipitation of ferric arsenate with higher insolubility than the calcium arsenate. Ferric ions, which are used as reducing agents, reduce the absolute values of the zeta potential of the particles leading to aggregation. The reaction to convert soluble arsenic acid to ferric arsenate occurs according to the following reaction:

$$Fe_2(SO_4)_3 + H_3AsO_4 => FeAsO_4 \downarrow (solid\ precipitate)$$

Ferric arsenate ($FeAsO_4 \cdot 2H_2O$) is a very stable arsenic compound. It is soluble, but the solubility decreases rapidly as the ratio of iron to arsenic increases. The solubility of ferric arsenate can be reduced several orders of magnitude if four to five times the stoichiometric amount of iron is present in a pH range of 3–7. At pH 4, the arsenic remaining in solution is about 8 mg/L at a Fe:As molar ratio of 1.5. Increasing the Fe:As ratio to 2 results in a significant reduction of arsenic in solution to about 0.15 mg/L. Further reduction of arsenic to about 0.02 mg/L can be obtained with a Fe:As ratio of 5.0. The data represented in the chart is experimental and represents near equilibrium conditions. Actual arsenic levels in operating systems will not achieve the same level of arsenic removal. We can see that at all Fe:As levels the minimum arsenic level in solution is obtained at about pH 4. Increasing the iron:arsenic mole ratio also results in a greater success in the solidification or stabilization of arsenic using iron.

Minimum solubility or maximum stability is ensured at a pH range of 3–7 when the iron:arsenic ratio is maintained at four times the stoichiometric ratio.

In several studies, an attempt has been made to simultaneously separate arsenic from water and to stabilize in the precipitate [25–31]. In the majority of cases, ferrous and ferric compounds are produced in such chemical precipitation processes. The presence of elements like Ca, Cd, Zn, Sr, Pb, Cu, and Mg have been reported to promote the stability of iron–arsenate precipitates, as the solubility of arsenic can be lowered significantly over a wide pH interval [32–34].

A mixture of iron (II), calcium hydroxide, and sulfuric acid can stabilize a high percentage of arsenic from aqueous arsenic waste as a matrix compound [35]. First of all, the arsenate ion gets reduced to an arsenite ion by the ferrous ion and then it forms insoluble ferric arsenate after iron air oxidation. After this a matrix compound ($Ca–Fe–AsO_4$ complex) is formed after addition of calcium hydroxide and sulfuric acid.

In the stabilization of arsenic rejects using iron oxides, ferrous sulfate, ferric sulfate, gypsum, and ferrous carbonate, it is found that ferric and ferrous sulfate provide over 95% stabilization efficiency [36,37].

A cement solidification study of arsenic–iron hydroxide sludge was reported [38] where arsenic was removed by coagulation with a ferric chloride compound. The calcium–arsenic matrix compound was produced by three mechanisms: (1) sorption onto C–S–H surfaces, (2) replacement of SO_4^{2-} by arsenic, and (3) reaction with cement components.

6.3.9 Reuse as construction materials

As environmental regulations become more stringent and volume of sludge generated continues to increase, traditional sludge disposal methods are coming under increasing pressure to change. Incineration is costly and contributes to air pollution, and landfill space is getting extremely limited. A possible long-term solution appears to be recycling of the sludge and using it for beneficial purposes.

Arsenic–iron sludge can be used in the brick manufacturing process [39]. The chemical composition of the arsenic treatment plant sludge (such as silica, alumina, and ferric oxide) is very close to brick clay composition. The physical and chemical properties of the produced bricks thus made should be determined and compared to conventional brick. The sludge proportion and firing temperature are found to be the two key factors in determining the quality of bricks. The compressive strength of prepared bricks initially increases and then decreases with the increase of sludge proportion. The optimum amount of arsenic sludge that upon mixing with clay produces strong brick is found to be 6% by weight. A typical brick manufacturing method using arsenic sludge and iron sludge is presented in the schematic

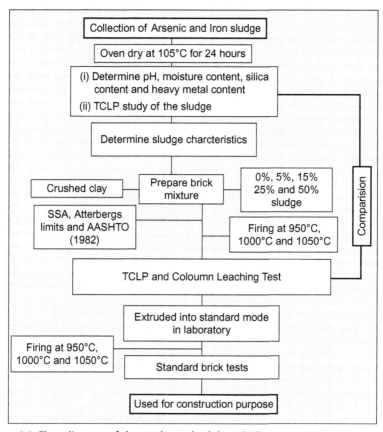

Figure 6.1 Flow diagram of the study methodology [40].

flow diagram in Figure 6.1 [40]. Water absorption capacity of brick is observed to decrease with the increase in firing temperature and decrease in the amount of sludge in the brick.

Mahzuz et al. [41] used arsenic-contaminated sludge in making ornamental bricks. They analyzed and justified the sludge effectiveness during the process of making ornamental brick. The detailed study was made upon the suitability of sludge in making bricks. Results of different tests specified that sludge proportion in the bricks' clay is the key factor for determining the quality of ornamental bricks. The result also shows that the compressive strength of ornamental bricks mutually decreases with an increase of sludge proportion and found an optimum value (4%) for making ornamental bricks. Above this limit, the quality of bricks or tiles may fall considerably. Different types of ornamental bricks, which are made in the Khadim Ceramic Industry for this research purpose, are shown in Figure 6.2.

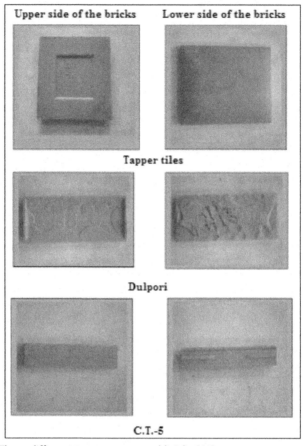

Figure 6.2 Three different types ornamental bricks [41].

6.3.10 Continuous removal of arsenic and stabilization: A novel approach

A membrane-integrated hybrid treatment system has been developed [42] for continuous removal of arsenic from contaminated groundwater with simultaneous stabilization of arsenic rejects for safe disposal. For the first time, an effective scheme for protection of the total environment has been ensured in this context where arsenic separated with a high degree of efficiency has been stabilized in a solid matrix of iron and calcium under response surface optimized conditions. Preoxidation of trivalent arsenic to pentavalent form was done by $KMnO_4$ in a continuous stirred tank reactor prior to nanofiltration of the contaminated groundwater that contained both trivalent as well as pentavalent arsenic. Nanofiltration investigations are

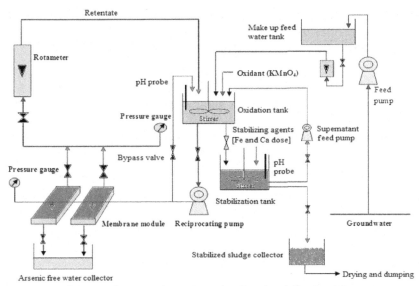

Figure 6.3 Schematic diagram of arsenic removal and stabilization [1].

carried out in a flat-sheet cross-flow membrane module well known for its capability of providing long service without significant fouling of the membrane surface. Figure 6.3 presents the flow sheet of the treatment plant.

In the continuous run, fresh make-up water (contaminated) is continuously added to the feed tank at the same rate at which treated arsenic-free water collected. After 5 months of continuous operation of a 5000 L/day water treatment plant, arsenic concentration in retentate side reaches a high level (25 mg/L). This means that after a period of 5 months, such a plant needs removal of the concentrated arsenic solution from the continuously operating loop. The next downstream operation is thus stabilization of this arsenic in a solid matrix, done through a coagulation–precipitation process. The system provides for periodic withdrawal of concentrated arsenic solution from the loop for precipitation and stabilization of arsenic rejects.

Response Surface Methodology of Design Expert Software (Version 8.0.6) may be adopted for arriving at the best optimization parameters of arsenic stabilization in a solid precipitation matrix in the coagulation–precipitation process. The TCLP tests for the solid arsenic matrix can be done to find the leaching out possibility. During the TCLP test, an extraction fluid is prepared by mixing with 5.7 mL glacial acetic acid and 64.3 mL of 1 N sodium hydroxide in 1 L deionized water and a maintained pH of 5 [43]. The fresh solid precipitate is mixed with the extraction fluid (20 times the solid weight). The mixing slurry is agitated for 18 hr at room temperature (25 °C). After the agitation sample is allowed 6–8 hr for settling, the extract is filtered through a filter paper of 0.45 μm pore size. The CWET [1] is also performed for assessing the stability of arsenic in the arsenic-bearing solid precipitate.

Around 140 L/m^{2}hr of water flux as well as over 98% of arsenic is removed during the nanofiltration process at a fixed 750 L/hr cross-flow rate, and 98% of arsenic stabilization is achieved at a ferric sulfate and calcium hydroxide dose of 250 mg/L and 500 mg/L, respectively, with a fixed pH of 5. The stabilization matrix successfully passes the TCLP test. The CWET can also be performed to further confirm stability of the matrix. FT-IR spectrum of the solid arsenic precipitate (Ca–Fe–AsO$_4$) indicates that there is no significant change in peak wavenumber before and after leaching tests of the arsenic-bearing precipitates.

6.4 CONCLUSION

Volumes of arsenic rejects from physicochemical treatment plants may be quite considerable, demanding special attention for safe disposal. Membrane-based treatment, however, does not produce such huge volumes of arsenic rejects and can be tackled safely even locally. Chemical coagulation-precipitation eventually leading to binding arsenic rejects in some solid matrix is the most practical solution and is most widely studied and suggested. Binding with some kind of cementing material has also been found to be very effective. Reuse with brick material appears to be another viable option. There has been limited research toward development of sustainable treatment technology that can safely manage and dispose of wastes generated by arsenic removal systems. Stabilization processes should be designed to address the needs of ultimate disposal. Stabilization of hazardous waste involves trapping the waste in a stable solid matrix, thus minimizing the escape of hazardous materials by leaching. This process also involves fixing or immobilizing the toxic elements by physical and or chemical means. A wide range of processes have been used in an attempt to successfully fix arsenic, and these processes include mixing the arsenic with various combinations of cement, lime, iron, silicates, and fly ash. It may be concluded that Portland cement with lime is appropriate for treating waste from sorptive filters but not oxidized precipitative sludges, where the high pH environment promotes desorption. Ferric hydroxides used to sorb arsenic in precipitative units retard cement set and in large quantity may destabilize cements owing to their differential expansion as humidity changes. This can result in a porous matrix with increased leachability. Optimum moisture content and waste:binder ratios depend on the chemical properties of the waste. There is evidence that Portland cement can immobilize soluble arsenites. Precipitation and solidification have been successfully used to stabilize arsenic-rich sludges and may be suitable for treating sludge generated by coagulation-precipitation. Difficulties in handling arsenic wastes are quite insignificant compared to the great relief that an arsenic treatment plant offers to the suffering people by providing them with safe drinking water.

REFERENCES

[1] Ghurye G, Younan JC, Chwirka J. Arsenic removal from industrial wastewater discharges and residuals management issues. J EUEC 2007;1.

[2] Yamauchi H, Fowler BA. Toxicity and metabolism of inorganic and methylated arsenicals. In: Nriagu JO, editor. Arsenic in the environment, part II: human health and ecosystem effects. New York: John Wiley & Sons; 1994. p. 35–53.

[3] Ashraf AM, Badruzzaman ABM, Jalil MA, Feroze AM, Kamruzzaman MD, Azizur Rahman M. Fate of arsenic in wastes generated from arsenic removal units. Dhaka: Department of Civil Engineering, Bangladesh University of Engineering and Technology; 2003. http://www.unu.edu/env/Arsenic/Dhaka2003.

[4] Hering JG, Chen P, Wilke JA, Elimelech M. Arsenic removal from drinking water during coagulation. J Environ Energ 1997;123:800–6.

[5] Baedeker MJ, Back W. Hydrogeological processes and chemical reactions at a landfill. Groundwater 1979;17:429–37.

[6] Fetter CW. Contaminant hydrogeology. USA: Prentice-Hall Inc.; 1999.

[7] Mudgal AK. A draft review of the household arsenic removal technology options. Rural Water Supply Network; 2001. http://www.htnweb.com.

[8] Das D, Chatterjee A, Samanta G, Roy Chowdhury T, Mandal BK, Dhar RK, et al. A simple household device to remove arsenic from groundwater and two years performance report of arsenic removal plant for treating ground water with community participation. In: Feroze Ahmed M, Ashraf Ali M, Adeel Z, editors. BUET-UNU international workshop on technologies for arsenic removal from drinking water, May 5–7, 2001, Dhaka, Bangladesh. p. 231–250.

[9] Singh TS, Pant KK. Solidification/stabilization of arsenic containing solid wastes using Portland cement, fly ash and polymeric materials. J Hazard Mater 2006;131:29–36.

[10] Leist M, Casey RJ, Caridi D. The fixation and leaching of cement stabilized arsenic. Waste Manag 2003;23:353–9.

[11] Sorini SS. Leaching tests: commonly used methods, examples of applications to coal combustion by-products and needs for the next generation. Laramie, Wyoming: Western Research Institute; 1997.

[12] Halim CE, Amal R, Beydoun D, Scott JA, Low G. Implications of the structure of cementations wastes containing Pb(II), Cd(II), As(V), and Cr(VI) on the leaching of metals. Cement Concr Res 2004;34:1093–102.

[13] Palfy P, Vircikova E, Molnar L. Processing of arsenic waste by precipitation and solidification. Waste Manag 1999;19:55–9.

[14] Mollah YA, Kesmez M, Cocke D. An X-ray diffraction and Fourier transform infrared spectroscopic investigation of the long-term effect on solidification/stabilization of arsenic(V) in Portland cement type-V. Sci Total Environ 2004;325:255–62.

[15] Akhter H, Cartledge FK, Amitava R, Tittlebaum ME. Solidification/stabilization of arsenic salts: effects of long cure times. J Hazard Mater 1997;52:247–64.

[16] Jing C, Liu S, Meng M. Arsenic leachability and speciation in cement immobilized water treatment sludge. Chemosphere 2005;59:1241–7.

[17] Kameswari KSB, Bhole AG, Paramasivam R. Evaluation of solidification/stabilization (S/S) process for the disposal of arsenic-bearing sludges in landfill sites. Environ Eng Sci 2001;18:167–76.

[18] Glasser FP. Fundamental aspects of cement solidification and stabilization. J Hazard Mater 1997;52:151–70.

[19] Twidwell LG, Mehta AK. Disposal of arsenic bearing copper smelter flue dust. Nucl Chem Waste Man 1985;5(4):297–303.

[20] De Villiers D. The preparation and leaching of arsenic doped slags [Thesis]. Melbourne, Australia: Monash University; 1995.

[21] Dutre V, Vandecasteele C. Solidification/stabilization of hazardous arsenic containing waste from a copper refining process. J Hazard Mater 1995;40:55–68.

[22] Dutre V, Vandecasteele C. Immobilization of arsenic in waste solidified using cement and lime. Environ Sci Technol 1998;32:2782–7.

[23] Shaw JK, Fathordoobadi S, Zelinski BJ, Ela WP, Saez AE. Stabilization of arsenic-bearing solid residuals in polymeric matrices. J Hazard Mater 2008;152:1115–21.

[24] Bankowski P, Zou L, Hodges R. Using inorganic polymer to reduce leach rates of metals from brown coal fly ash. Miner Eng 2004;17:159–66.

[25] Gulledge J, O'Connor J. Removal of arsenic (V) from water by adsorption on aluminium and ferric hydroxide. J Am Water Works Assoc 1973;65(8):548–52.

[26] Roberts LC, Hug SJ, Ruettimann T, Khan AW, Rahman MT. Arsenic removal with iron (II) and iron (III) in waters with high silicate and phosphate concentrations. Environ Sci Technol 2004;38:307–15.

[27] Baskan MB, Pala A. Determination of arsenic removal efficiency by ferric ions using response surface methodology. J Hazard Mater 2009;166:796–801.

[28] De Klerk RJ, Jia Y, Daenzer R, Gomez MA, Demopoulos GP. Continuous circuit co-precipitation of arsenic(V) with ferric iron by lime neutralization: process parameter effects on arsenic removal and precipitate quality. Hydrometallurgy 2012;111–112:65–72.

[29] Daenzer R, Xu L, Doerfelt C, Jia Y, Demopoulos GP. Precipitation behaviour of As(V) during neutralization of acidic Fe(II) − As(V) solutions in batch and continuous modes. Hydrometallurgy 2014;146:40–7.

[30] Robins RG. The stability and solubility of ferric arsenate: an update. In: Gaskell DR, editor. EPD congress '90. Warrendale PA: TMS; 1990. p. 93–104.

[31] Bluteau MC, Demopoulos GP. The incongruent dissolution of scorodite solubility, kinetics and mechanism. Hydrometallurgy 2007;87:163–77.

[32] Harris GB, Monette S. Productivity and technology in the metallurgical industries. Germany: TMS-AIME/GDMB Joint Symposium; 1989.

[33] Emmett MT, Khoe GH. Extraction and processing division congress. TMS Publications; 1994.

[34] Khoe G, Carter M, Emmett M, Vance ER, Zaw M. 6th AusIMM extractive metallurgy conference, Australia; 1994, p. 281–286.

[35] Sandesara MD. Process for disposal of arsenic wastes. U.S. Patent 4,118,243; 1978.

[36] Artiola JF, Zabcik D, Sidney HJ. In situ treatment of arsenic contaminated soil from a hazardous industrial site: laboratory studies. Waste Manag 1990;10(1):73–8.

[37] Song S, Lopez-Valdivieso A, Hernandez-Campos DJ, Peng C, Monroy-Fernandez MG, Razo-Soto I. Arsenic removal from high-arsenic water by enhanced coagulation with ferric ions and coarse calcite. Water Res 2006;40:364–72.

[38] Phenrat T, Marhaba TF, Rachakornkij MA. SEM and X-ray study for investigation of solidified/stabilized arsenic–iron hydroxide sludge. J Hazard Mater 2005;118:185–95.

[39] Hassan KM, Fukushi K, Turikuzzaman K, Moniruzzaman SM. Effects of using arsenic–iron sludge wastes in brick making. Waste Manag 2013; http://dx.doi.org/10.1016/j.wasman.2013.09.022.

[40] Rouf MA, Hossain MD. Effects of using arsenic-iron sludge in brick making. In: Fate of arsenic in the environment, Proceedings of the BUET-UNU international symposium, Dhaka, Bangladesh, February 5–6; 2003.

[41] Mahzuz HMA, Alam R, Alam MN, Basak R, Islam Use MS. of arsenic contaminated sludge in making ornamental bricks. Int J Environ Sci Technol 2009;6(2):291–8.

[42] Pal P, Chakrabortty S, Linnanen L. A nanofiltration–coagulation integrated system for separation and stabilization of arsenic from groundwater. Sci Total Environ 2014;476–477:601–10.

[43] Environmental Protection Agency US. 40 code of regulations; 1992 (Part 261.31).

CHAPTER 7

Arsenic Removal Technologies on Comparison Scale and Sustainability Issues

Contents

7.1 INTRODUCTION

The vastness of the arsenic contamination problem requires an all-out effort to combat such a problem. Due to the very insidious nature of the problem and the absence of effective monitoring in many cases, it is difficult to assess exactly how many people daily are joining the long list of arsenic victims. To ensure sustainable supplies of safe drinking water to the arsenic-affected areas, certain basic requirements must be fulfilled.

1. An adequate supply source has to be identified and the best source for a specific area has to be selected.
2. Efficient treatment systems should be developed for treating such water to at least the WHO-prescribed level of 10 ppb.

Groundwater Arsenic Remediation
http://dx.doi.org/10.1016/B978-0-12-801281-9.00007-2

3. Treatment cost of such systems should be affordable to the people of the affected regions.
4. Long-term operation of such treatment systems should be ensured through effective monitoring and regular maintenance.

A sustainable system in the present context is expected to ensure long-term trouble-free operation, yielding safe drinking water at a reasonably low price. Such a system should be able to take care of the total environment without transferring the problem of pollution from one area to another. The system should be considered in the context of ground realities of the affected people, where ease of maintenance, low price, and high quality of water are of paramount importance. Ease of maintenance, simplicity, and flexibility of the system are other important parameters that should be considered when identifying a sustainable system. Before providing a detailed comparison of the available systems, a general consideration of the major merits and demerits of such systems will help in choosing a specific process or technology.

7.2 MERITS AND DEMERITS ASSOCIATED WITH BROAD ARSENIC ABATEMENT SYSTEMS

7.2.1 Membrane-based systems

Membranes have proven to be most effective in producing almost arsenic-free water. A major advantage in using membranes for water purification is that membranes remove not only arsenic, but other water contaminants such as chromium and iron also. Nanofiltration (NF) and reverse osmosis (RO) can remove viruses, bacteria, and almost all types of water contaminants. The most refractory and challenging contaminants can also be removed.

Membrane distillation produces water with the highest purity but suffers from the disadvantage of low volumetric flux. Membrane-based processes are known for a high degree of separation and low flux over long-time operation when fouling is significant. Widely studied modules like hollow fiber, spiral wound, plate and frame, and tubular types are largely fouling-prone, necessitating frequent replacement of membranes. Among the membrane-based processes, the nano-membrane-based flat-sheet, cross-flow filtration module stands out as one of the most promising because it has the potential of ensuring high rejection of arsenic and high flux of pure water in a largely fouling-free membrane module at a reasonably low transmembrane pressure compared to that required in RO. The toxic compounds can be separated by NF membranes exploiting both steric (sieving) and Donnan (electrical) mechanisms, depending on the characteristics of such compounds and the membranes involved.

Both NF and RO can very effectively remove arsenic from drinking water. Removal efficiency can reach over 99%. But NF can even outperform RO in terms of higher flux for the same rate of removal of arsenic while permitting filtration at a lower pressure than is necessary for RO. Low pressure operation is again a great advantage *vis-à-vis* RO types in the affected rural areas where power supply is a major constraint. Though most of the NF studies show that removal efficiency for As(V) is higher than that for As(III), wide variations have been reported with respect to the extent of such separation. That As(V) removal is much better than As(III) removal points to the necessity of the use of the oxidation technique prior to membrane filtration if the presence of As(III) in water is significant. Nanofiltration in such a region can be the most promising technology [1]. Detailed modeling and at least preliminary economic analysis for such systems have been done [2]. However, to raise scale up confidence further, field level installation of units is necessary. RO and NF membranes still remain expensive in several arsenic-affected regions, particularly in South East Asia. However, the relatively high cost of membranes can very effectively be offset by the low cost of operation of simple plants. The advantages of using membrane in arsenic removal are (1) the membrane technologies can effectively remove portions of all dissolved solids including arsenic from feed water and even prevent the microorganisms passing through the membrane to diminish the harmful diseases and (2) the membrane itself does not accumulate arsenic, so disposal of used membranes would be simple, maintenance and operation requirements are minimal, and no chemicals (except oxidizing agents in the case of microfiltration, ultrafiltration, nanofiltration, reverse osmosis, and electrodialysis) need to be added. The only hurdle in the developing countries may be supply of membranes at a low price.

7.2.2 Adsorption-based process

Adsorption-based purification of water has been very widely studied. Several adsorbents have been examined for assessing effectiveness of arsenic separation from water on a small scale; some have been developed [3] for removal of not only arsenic from water but also other contaminants. Adsorption-based units are best suited as community water filters. A relatively simple system, an adsorption-based unit initially separates the contaminants with a high degree of efficiency since the effective driving force for mass transfer is high at the beginning. However with time, as the adsorption bed gets more and more saturated and exhausted, it loses its capacity of separation and eventually no further separation is done. From a quality point of view, because the adsorption-based process is not self-monitoring, it continues to produce water even after the adsorption bed gets exhausted. Often villagers in the affected areas are misled as they continue to drink water collected from the adsorption plant that is really not arsenic free. In remote areas, without continuous monitoring of the quality of treated water through chemical or instrumental analysis, adsorption units may often be misleading. Periodic replacement of adsorbent material is a must for such units. Heavy deposition of iron on the adsorption bed often clogs the flow channels in an adsorption bed, turning the unit defunct, as is experienced in many areas of the Bengal–Delta basin. Regeneration of spent adsorbent is energy-intensive, often incomplete, and involves high cost.

7.2.3 Chemical coagulation–precipitation

Physico-chemical treatment can produce treated water in large volumes, but the degree of purification cannot be expected to be high or even comparable to those obtained in membrane separation. Disposal of a huge amount of sludge is another problem. For large-scale treatment of arsenic-contaminated groundwater in the arsenic-affected areas of developing countries, there is hardly any alternative to physico-chemical coagulation–precipitation of arsenic from drinking water. Particularly where the river is far away from affected villages, this low-cost technology is likely to be the most promising.

Physico-chemical separation through chemical coagulation and precipitation has been demonstrated by many researchers [4,5] as one of the most effective methods of arsenic separation. Such studies have considered a number of combinations of coagulants and oxidants with a considered effect on pH. The most widely used oxidant is $KMnO_4$ and the most widely used coagulant is ferric chloride. Major advantages are that it permits large-scale purification at a relatively low cost, but separation efficiency remains low at 90–92%. Coagulation–precipitation methods are subclassified based on the use of oxidants. Efficiency and cost of treatment also vary according to the use of such oxidants. Such physico-chemical treatment plants may also be installed for treating surface water as an alternative to groundwater where groundwater is arsenic-contaminated. However, in such cases the plant will treat surface water for removal of other organic or inorganic substances, but not for removal of arsenic or other metallic contaminants that may be present in surface water due to discharge of industrial or municipal wastewater.

7.2.3.1 Performance and limitation of arsenic oxidants

Oxidants such as ozone, hydrogen peroxide, chlorine, and permanganate in arsenic removal processes are used mainly for changing oxidation states of As(III) to As(V). Their performance and limitations are discussed individually as follows.

7.2.3.2 Ozone (O_3)

Ozone may be the most satisfactory preoxidation method for converting As(III) to As(V) in water along with disinfecting harmful bacteria and/or pollutants. It generally leaves behind no by-products. Ozone, when added to water containing arsenic and soluble iron, will oxidize both arsenic and iron, forming sites on the ferric hydroxide for arsenic to adsorb. The arsenic-bearing iron hydroxide can then be removed by solid liquid separation processes. Ozone preoxidation before nanofiltration could present a problem if the Assimilable Organic Carbon (AOC) that is formed has a low molecular weight and passes through the membrane.

7.2.3.3 Hydrogen peroxide (H_2O_2)

Hydrogen peroxide oxidation was effective but limited by reactions with calcium hydroxide. After oxidation, the resulting arsenate waste was effectively stabilized using ferric sulfate.

7.2.3.4 Chlorine (Cl_2)

Chlorine is a good oxidant for As(III), but application must come early in the treatment train when disinfectant by-product precursor concentration is high and there is a danger of producing large concentrations of disinfectant by-products.

7.2.3.5 Potassium permanganate ($KMnO_4$)

Potassium permanganate may work better than chlorine, however, there is no sufficient information on the permanganate demand for arsenic oxidation relative to the demand exerted by other substances.

7.3 COMPARISON OF TECHNOLOGIES IN TERMS OF ARSENIC REMOVAL EFFICIENCY

Arsenic removal efficiency of different processes can be compared from literature [6,7] as presented in Table 7.1.

Table 7.1 Percentage Removal Figures Are for As(V) Removal

Treatment process	Maximum removal %
Alum precipitation process	90
Iron precipitation	95
Lime softening process (pH >10.5)	90
Coprecipitation	>90
Ion exchange (sulfate 50 ppm)	95
Activated alumina	95
Reverse osmosis, nanofiltration, electrodialysis	>98
Membrane distillation	>99

7.4 COMPARISON OF RESIDUAL GENERATION AND DISPOSAL METHODS

All physico-chemical and membrane processes in arsenic removal described above differ in residual production and residual management options. Table 7.2 describes the type of residual produced and a list of possible disposal methods for the residual.

7.5 OVERALL QUALITATIVE COMPARISON

The comparison is subjective in nature because no hard figures are used. The comparison is based on several parameters. Ten arsenic removal processes are compared in Table 7.3.

Chemical intensity is a measure of the quantities and numbers of the chemicals used by a particular process whereas power intensity stands for relative power consumption of the processes. Labor intensity is a measure of the need for operator attention on the process. Area required refers to the land area required for a plant. Waste is based on whether or not solid and/or liquid wastes are produced by the processes. Removal efficiency refers to an indication of how much of the arsenic in feed water is removed by a process. Well site adaptability ratings are an indication of the suitability of a process for wellhead treatment.

7.6 TOWARD SUSTAINABLE SOLUTIONS

7.6.1 Selection of sustainable technology

The three major approaches toward sustainable solutions are (1) better supply side management, (2) better demand side management, and (3) adoption

Table 7.2 Residual Generation and Disposal for the Various Treatment Methods

Treatment Method for arsenic	Form of residual	Residual generation	Disposal
Ion exchange	Liquid	Regeneration streams Spent backwash Spent regenerant Spent rinse stream	Sanitary sewer Discharge Evaporation ponds/lagoon
	Solid	Spent resins	Landfill Hazardous waste landfill Return to vendor
Adsorption	Liquid	Regeneration streams Spent regenerant Spent rinse stream Spent backwash	Sanitary sewer Direct discharge Evaporation ponds/lagoon
Coprecipitation	Liquid	Filter backwash	Sanitary sewer Direct discharge Evaporation ponds/lagoon
	Solid	Sludge (if separated from backwash water) Spent media	Sanitary sewer Land application Landfill Hazardous waste landfill
Membrane Processes	Liquid	Brine (reject and backwash streams)	Direct discharge Sanitary sewer Deep well injection Evaporation ponds/lagoon

Sources: References [8,9].

of appropriate water treatment technology. To arrive at the most sustainable solution out of a host of apparently possible solutions [10], we have to first analyze the root causes of the problem.

The major sources of arsenic in water are anthropogenic, geogenic, coal mining–related activities, petroleum mining–related activities, volcanogenic activities, and industrial activities. Anthropogenic arsenic is caused by human activities. This results from the use of a variety of arsenic compounds in various applications such as production of agrochemicals, wood preservatives, nonferrous alloys, electronic goods, chemical munitions, and metal extraction from

Table 7.3 Overall Comparison of Arsenic Removal Processes

	Precipitation processes				Membrane processes						Adsorption Processes	
Comparison	Alum	Iron	Lime	Coprecipitation	MF	UF	NF	RO	SD-MD	ED	AA	IE
Intensity												
Chemical	H	H	H	H	L	VL	VL	VL	N	VL	M	L
Power	M	M	M	M	M	M	H	H	M	H	L	L
Labor	H	H	H	H	L	L	L	L	L	L	L	L
Area required	H	H	H	H	L	L	L	L	L	L	L	L
Waste												
Solid	Y	Y	Y	Y	VL	N	N	N	N	N	N	N
Liquid	N	N	N	N	Y	Y	Y	Y	Y	Y	Y	Y
Removal efficiency	L	L	L	L	L	L	H	H	VH	H	H	H
Well site adaptability	L	L	L	L	H	H	H	H	H	H	H	H

arsenic-bearing mineral ores. Geogenic source is represented by arsenic in coal-bearing units that naturally get released into groundwater during natural weathering. Release of arsenic from iron oxide is another abundant source of geogenic arsenic. High concentrations of arsenic in aquifers cause its absorption by the surface of iron oxides, which again gets released in the reducing environment in the presence of microbial activities and anaerobic conditions.

Mining activities and petroleum extraction activities hugely contribute to dissolution of arsenic from the minerals to groundwater. Mine tailings with high concentration of arsenic have caused contamination of both surface water and groundwater in many areas; the Khetri mine area in Rajasthan in India is one such example. Mining of groundwater or excessive withdrawal of groundwater in Bangladesh and West Bengal in India is believed to have caused severe geological disturbance, exposing arsenic-bearing mineral to atmospheric oxygen, causing it to leach out. What transpires from these observations is that the problem of arsenic in groundwater has its origin in several natural phenomena as well as many human activities, ranging from agricultural activities that withdraw huge amounts of groundwater to mining and industrial activities that withdraw minerals from the earth.

Related to these are a host of issues ranging from consumption pattern, to life style, to economic development. Thus rise in population, rise in massive demand for water, urbanization, industrialization, economic development, lifestyle, and consumption pattern all are related to the problem of arsenic contamination. In a holistic approach such issues need to be looked at for the sustainable development of water resources. The best technology or the most sustainable technology for supplying safe drinking water to the people of the affected regions across the world will vary from place to place. Despite such possibilities of variation, a simple nanofiltration system in flat-sheet cross-flow modules appears to be quite sustainable.

Unless concepts of sustainable development spread across societies and countries, the issue of arsenic contamination cannot be tackled successfully. Let us now examine how we can choose better supply side management and demand side management of water apart from the installation of sustainable systems for purifying contaminated water.

7.6.2 Selection and adoption of sustainable water management strategy

As geological disturbance resulting from overwithdrawal of groundwater has largely been held responsible for arsenic contamination of groundwater in many areas, remedial measures directed toward reducing pressure of such withdrawal are likely to be successful. Therefore, it is suggested that the remedy should be sought through better management of surface water so that geological disturbance of the underground aquifers can be minimized in a regime of controlled groundwater use. As the major portion of the groundwater in the Bengal delta basin is used in rice-dominated, highly water-intensive agriculture, development of surface water-based irrigation can effectively control withdrawal of groundwater. Region-specific solutions through better supply and demand side management and through institutional mechanisms can be found in Roy et al. [11].

The obvious choice for the best treatment is the nanofiltration-integrated hybrid treatment plant as detailed in Chapter 4. However, only a one-time commissioning of such a plant will not ensure a sustained supply of safe drinking water. Such a plant needs to be managed with involvement of the stakeholders. A contingent valuation approach may help arrive at local level management strategy of the treatment plant [12].

Ultimately lack of access to safe drinking water for a section of the population reflects on poor economic conditions of the affected people, lack of physical planning, and lack of development of alternate water resources. This demands an integrated approach to development, management, and use of water in general. While scientists and technologists may continue to do their research toward more cost-effective, efficient, and innovative technologies to give relief to the people of arsenic-affected regions, policymakers may learn from the examples of the lowest per capita loss of water in Denmark and the best physical planning of Sweden, where people enjoy high environmental quality. Use of economic instruments may also significantly contribute to sustainable water resource development and management. Concepts of sustainable development should be institutionalized across countries. From educational institutes to every corner of society, sustainable development concepts should guide people in developing a sustainable society that will automatically initiate actions for sustainable town and village-level planning, and integrated water resources development, management, and use.

REFERENCES

[1] Pal P, Chakrabortty S, Linnanen L. A nanofiltration-coagulation integrated system for separation and stabilization of arsenic from groundwater. Sci Total Environ 2014;476–477:601–10.

[2] Pal P, Chakraborty S, Roy M. Arsenic separation by a membrane-integrated hybrid treatment system: modeling, simulation and techno-economic evaluation. Sep Sci Technol 2012;47–8:1091–101.

[3] Sen M, Pal P. Treatment of arsenic-contaminated groundwater by a low cost activated alumina adsorbent prepared by partial thermal dehydration. Desalin Water Treat 2009;11:275–82.

[4] Hering G. Arsenic removal from drinking water during coagulation. J Environ Eng 1997;800–7.

[5] Pal P, Ahammad Z, Pattanayek A, Bhattacharya P. Removal of arsenic from drinking water by chemical precipitation—a modelling and simulation study of the physical-chemical processes. Water Environ Res 2007;79(4):357–66.

[6] Pal P, Sen M, Manna AK, Pal J, Roy SK, Roy P. Contamination of groundwater by arsenic: a review of occurrence, causes, impacts, remedies and membrane-based purifications. J Integr Environ Sci 2009;6:1–22.

[7] Macedonio F, Drioli E. Pressure-driven membrane operations and membrane distillation technology integration for water purification. Desalination 2008;223:396–409.

[8] Choong T, Chuah TG, Robiah Y, Koay FLG, Azni I. Arsenic toxicity, health hazards and removal techniques from water: an overview. Desalination 2007;217:139–66.

[9] Sorg TJ. Regulations on the disposal of arsenic residuals from drinking water treatment plants. EPA Contract 68-C7-0011, Work Assignment 0-38, EPA/600/R-00/025; 2000.

[10] Kartinen EO, Martin CJ. An overview of arsenic removal processes. Desalination 1995;103:79–88.

[11] Roy M, Nilson L, Pal P. Development of groundwater resources in a region with high population density: a study of environmental sustainability. Environ Sci 2008; 5(4):251–67.

[12] Roy M, Chakrabortty S. Developing a sustainable water resource management strategy for a fluoride-affected area: a contingent valuation approach. J Clean Techn Environ Pol 2014;16:341–9.

INDEX

Note: Page numbers followed by *f* indicate figures and *t* indicate tables.

Printed in the United States
By Bookmasters